O9-BTJ-615

❧ PRAISE FOR ❧
MAKE MEAD LIKE A VIKING

"A great guide to mead making, full of practical information and fascinating lore."

—SANDOR ELLIX KATZ, author
of *The Art of Fermentation* and *Wild Fermentation*

"Jereme Zimmerman has captured the wild spirit of mead quite literally—as the quintessential naturally fermented beverage of humankind from the beginning, which reached its apotheosis with the Vikings. Without compromising its mysterious allure, he brings it down to earth for all to make and enjoy."

—PATRICK E. MCGOVERN,
author of *Ancient Wine* and *Uncorking the Past*

"This is a fun book—and fortunately, it doesn't stop there. Coupled with the fun parts is a book that is informative and detailed in everything from choosing honey all the way to what kinds of corks to use. As a beekeeper who has always had lots of good raw honey on hand, I have made mead before but only in the kind of sterile environment that Jereme Zimmerman eschews. His book opened my eyes to the possibility of returning to the much more natural and time-honored ways of brewing this fascinating beverage."

—JEFFREY HAMELMAN,
director, King Arthur Flour Bakery;
author of *Bread: A Baker's Book of Techniques and Recipes*

"*Make Mead Like a Viking* puts the ME back in mead: my Scandinavian heritage simply sang when reliving the history, reading the recipes, and playing the drinking games he includes. And best yet . . . Zimmerman encourages mead makers to keep their own bees! There's no better way to get the best honey there is than when you, and the bees you care for, make it happen together. For me, this is the perfect marriage."

—KIM FLOTTUM, editor-in-chief,
Bee Culture: The Magazine of American Beekeeping

"I really delighted in this inspired and informative read. Throw caution into the mead-making wind and relish the challenge of some of the more unusual flavorings and ingredients. I now feel more like being a Viking mead maker than ever, and coming from a Celt and fourth-generation mead maker that is something! Enjoy mead and make merry men and maidens."

—SOPHIA FENTON, director,
Cornish Mead Co. Ltd.

"Tradition meets modernity in this marvelous look at the ancient brewing of honey-based beverages."

—MIKE FAUL, owner and brewmaster,
Rabbit's Foot Meadery

Make Mead Like a Viking

Traditional Techniques for Brewing Natural, Wild-Fermented, Honey-Based Wines and Beers

Jereme Zimmerman

Chelsea Green Publishing
White River Junction, Vermont

Copyright © 2015 by Jereme Zimmerman.
All rights reserved.

Unless otherwise noted, all photographs
by Jereme Zimmerman.

No part of this book may be transmitted or reproduced in any form by any means without permission in
writing from the publisher.

Project Manager: Patricia Stone
Editor: Michael Metivier
Copy Editor: Laura Jorstad
Proofreader: Ben Gleason
Indexer: Shana Milkie
Designer: Melissa Jacobson

Printed in the United States of America.
First printing October, 2015
10 9 8   21 22

Our Commitment to Green Publishing

Chelsea Green sees publishing as a tool for cultural change and ecological stewardship. We strive to align
our book manufacturing practices with our editorial mission and to reduce the impact of our business
enterprise on the environment. We print our books and catalogs on chlorine-free recycled paper, using
vegetable-based inks whenever possible. This book may cost slightly more because it was printed on paper
from responsibly managed forests, and we hope you'll agree that it's worth it. *Make Mead Like a Viking* was
printed on paper supplied by Versa Press that is certified by the Forest Stewardship Council.

Library of Congress Cataloging-in-Publication Data

Zimmerman, Jereme, 1976–
  Make mead like a viking : traditional techniques for brewing natural, wild-fermented, honey-based wines
and beers/Jereme Zimmerman.
    pages cm
  ISBN 978-1-60358-598-9 (paperback) -- ISBN 978-1-60358-599-6 (ebook)
1. Mead. I. Title.

  TP588.M4Z56 2015
  663'.4—dc23

                      2015023034

Chelsea Green Publishing
85 North Main Street, Suite 120
White River Junction, VT 05001
(802) 295-6300
www.chelseagreen.com

To my lovely, ever-encouraging wife, Jenna Lee, and our precocious daughters, Sadie Lee and Maisie Jane—I offer my eternal love and gratitude . . . and my sincere apologies for having to put up with me while writing this book.

And to my father, Wayne "Waldo" Zimmerman, and mother, Janice Zimmerman—my tendencies toward self-sufficiency, simple living, and natural brewing are due to the solid foundation you provided. I owe you some farmwork, a debt that I have no time to pay. It's the thought that counts, right?

And finally, to the dynamic trio of Vikings and mead: Dave, Zach, and Steve—may you enjoy bountiful mead from the teats of Heidrun the goat for eternity in the halls of Valhalla.

❋ CONTENTS ❋

❄ RECIPE LIST ❄

❧ PREFACE ❧

When I first started on the path to brewing like a Viking, my intent was simple—to understand how the Norse and other ancient cultures got by without all the gadgets, chemicals, laboratory-bred yeast strains, and overly complex techniques that seem to pervade the modern mind-set with regard to homebrewing.

Having grown up on a Kentucky goat farm—and returning to my homesteading roots later in life—I am accustomed to living frugally and eating and drinking primarily what my family produces by working the land. My dad made wine using only ingredients he grew or got free from friends, which is a very Viking way of doing things (as long as you change "got free from friends" to "obtained by bartering with neighboring farmsteads and plundering distant lands"). The Norse had to live through long, cold winters and short summers, so they had to preserve as much of their food as possible. Since fermentation is a natural preservative—and brewing and drinking mind-altering beverages is a good way to pass the time when cramped up in a hut in the dead of winter—making mead and other fermented beverages was integral to survival in the dark, cold wilderness of ancient Northern Europe. In the increasingly dark wilderness of modern life, where we are constantly inundated with mass-produced, homogenized products saturated in chemicals and GMOs (genetically modified organisms), we can choose to emulate the Norse and other ancient cultures by bringing a sense of wildness, mysticism, and individuality to our unique, home-crafted brews.

I'll admit that it took me some time to reach this mind-set. When I first got into homebrewing, I brewed beer. I had just moved to Seattle, a mecca for craft-beer lovers. After sampling my share of the many local brews available, I decided I needed a new hobby. While living in an apartment on North 85th in the Greenwood district, I hopped a bus to the homebrewing store The Cellar, where I chatted with the staff about the essentials I would need to start brewing beer. I chose to go with a basic beer-brewing kit, a copy of Charlie Papazian's *The Complete Joy of Home Brewing*, and the ingredients for a raspberry wheat.

My first batch turned out to be quite tasty—although, looking back, it tasted more like a raspberry soda. I haven't brewed using fruit extract since, always choosing to go with whole fruit. As I continued brewing, my brews quickly became more experimental. Experimentation, after all, is what homebrewing is all about. As much as I appreciate what I learned from using those early beer kits, I now find myself wishing I'd focused less on the extreme sanitization, strict temperature parameters, and other regimented guidelines espoused by homebrewing manuals.

Over time, I began to garner an interest in brewing with natural, organic processes, and started to think on how people would have made alcohol before all the fancy equipment and brewing kits were available. Concurrently, I started trying various commercial meads with my friends (and fellow mead makers) Dave Brown and Josh Parker in the "Dave Cave," and my brother Zach in the Appalachian Viking mead hall he built on the back 40 of our parents' farm. The wheels were turning, but I wasn't quite sure yet where they were heading.

My friend Dan Adams, a computer programmer by vocation (and my cousin Leah's husband), had been talking about building an online homesteading network that would connect people locally to trade farm and garden goods. I figured this was just going to be a part-time hobby for him, and that I would likely participate to some degree. But it became bigger than that—quickly becoming a mecca for fervent free-thinkers and homesteaders, and attracting a community of enthusiastic individualists who trade goods, ideas, passions, and lofty goals—all while indulging in good humor and camaraderie.

Needless to say, Earthineer became a full-time job for Dan. During its early years, he asked me to contribute. I wasn't sure what I had to offer, but figured I'd at least create a profile. Seeing as my hair (mostly my beard these days, which is becoming increasingly white) has a reddish tint to it, I decided on the online handle RedHeadedYeti. There was, after all, already a GnomeNose, a GrumpyOldMan, and a BickensChickens on the site, so I figured I'd be in good company. One of my first "Yeti" articles for Earthineer was titled "Mead Making:

Techniques Old & New," and was essentially my announcement to the world that I was beginning my journey into learning the brewing secrets of the ancients.

A depiction of the author brewing a batch of ancient, mystical mead, by Michael Startzman.

With each subsequent article, I recounted what I continued to learn from my research, experimentation, and discussions with beer brewers, mead makers, and vintners from all corners of Midgard. My techniques changed, growing more ancient and wild. By modern reasoning, what I was doing was highly experimental. However, in my mind, I was doing nothing new, but rather rediscovering well-established practices that have been largely tossed aside by modern society in its nearsighted march toward "progress." The practices I was learning about, and will discuss in this book, were an integral part of life for much longer than what has become the standard today. I'm all for taking advantage of modern conveniences, but I have a tendency, like one-eyed Odin, to keep one eye focused on what surrounds me and the other closed to allow contemplation on what came before and what is within. Because of this, I have a tendency to run into things and trip over my own feet, but I often learn things about myself and the world around me in the process.

Initially, I assumed this obsession with brewing like the ancients was just an odd interest held by myself and a few friends, but as I began presenting workshops and writing more articles, the feedback I received made it clear that I was on to something. Hence, you have this book—an extended foray into the journey I started in my web and magazine articles, and workshops. I've worked to present the practical details on mead making as clearly as possible without getting bogged down in boring (and largely unnecessary) technical lingo, and have incorporated where appropriate the deep historical and mythological origins of this vastly enjoyable pastime. I hope you have as much fun with it as I did.

Skål!

❉ INTRODUCTION ❉

Mead. Vikings. For many, it's impossible to think of one without the other. Go ahead, close your eyes, clear your mind, and try. As a matter of fact, before proceeding any further, put this book down, find a quiet spot, and take a moment to sit quietly with your eyes closed and forget everything you know. Forget what you've been told about brewing alcoholic beverages—that it requires all sorts of fancy equipment and laboratory-produced yeasts and chemicals, and that it's a complicated, time-consuming process. Drop all your notions about the current state of food and drink production in a society that has become increasingly obsessed with uniformity and an overemphasis on sanitization. Take note, though, that I'm not talking about not being clean in producing your own food and drink. While the oversterilization

An Appalachian Viking mead hall built by the author's brother, Zachary, on the back 40 of their childhood home, a goat farm in Kentucky.

prevalent in commercial production is ultimately necessary due to the strong potential for contamination, it also results in killing off all those good microbes that bring life to fermented foods and drinks.

As you're coming closer to a state of Viking Zen, allow your mind to detach itself from your body and travel back to an ancient world of fire and ice in which mythology, magic, and reality were so intertwined as to be one and the same. Now that you're settled comfortably in the land of the ancients, raise your hand and point one finger skyward toward Asgard, the home of the Norse gods and fallen Vikings. Open one eye to emulate Odin, who drank from the well of Mimir to gain wisdom, leaving behind one of his eyes. Stand up slowly and start spinning. Spin faster. Dance with wild abandon! Go berserk! Imagine yourself as a Viking decked out for battle, having just fortified yourself with a hearty meal of roast goat and mead. If you've already done so, even better. If not, perhaps this has helped you feel a bit intoxicated . . . and free. Free to think, brew, and create, unfettered by the chains of a post-industrial, post-Pasteur world.

With your load now lightened a bit, prepare an environment conducive to inviting in the *bryggjemann* and brew pixies. Don't worry if you don't know quite what I'm talking about. All will be made clear in time. Even though you won't be brewing just yet, the goal here is to ensure that when you do begin, you've prepared the proper setting for brewing like a Viking. If you are into folk, medieval, or Viking-themed music of any sort, put together a playlist of your favorites. If not, this is a good time to start. For the Norse and many other ancient cultures, music, dance, meditation, and community were all integral components of the brewing process. To them, this was magic. They didn't call upon the Internet or head to the local brewing store to choose a yeast that had been prepared for them by men (or women) in white coats. They called upon the gods, quite literally as a matter of fact. Ancient mead-making traditions extend far beyond Nordic lands—Ethiopia, India, China, indigenous South American tribes . . . the list goes on. Nearly every culture has a deep connection with honey and the fermented beverages that can be made from it. In order to return to the simpler practices of our varied ancestral milieu, it is necessary that we unlearn complex and expensive modern methods of making honey-based ferments to create beverages that are truly magical—and healthful.

This is not a typical brewing manual—but you probably already guessed that from the title. Many of the techniques go against what you will read in modern brewing books that dwell heavily on tools, equipment, and ingredients that wouldn't have been available to ancestral cultures. The techniques I will outline have been tested thoroughly by myself and other DIY fermentation enthusiasts who have been gracious enough to share their processes with me. Not every brew you make will turn out perfect, but don't let that stop you. Don't feel that you need to go to a homebrewing store or visit a homebrewing website before you can get started in homebrewing, either. You *can* make excellent mead, beer, and wine using primarily what you find in your garden, from local farms and beekeepers and—by extension—farmers markets and co-ops, and in the wildness of nature.

Nor is this a book that will consist primarily of straightforward brewing recipes. While I have interspersed recipes and outlined techniques throughout the text, my goal is to weave a narrative on *why* you should brew (and drink) like the ancients. Research into ancient mindsets pertaining to the brewing of alcohol, storytelling, and mythmaking are the prevalent notions that fueled this book. Alcohol was looked at very differently in the distant—and even recent—past, and we will thus be transporting ourselves back in time so that we can approach brewing with a distinctively pre-modern mentality. In making wildcrafted brews, you will learn that it's more important to develop a rhythm than to follow strict guidelines. Before you know it, you'll have shelves full of burbling airlocks and living ferments. Be warned—making your own ferments can become an addiction. Not the negative kind with all the modern societal implications, but a healthy, enlivening, and—we can only hope—world-changing addiction.

So how about it? Are you ready to head into the ancient, magical past and learn to make mead like a Viking?

Asgardsreien by Norwegian historical painter Peter Nicolai Arbo (1831–92) depicts the Wild Hunt (or the Hunt of Odin), a Norse mythology motif that heralded war or other cataclysms. It can also be found in the folklore of many other ancient Northern, Western, and Eastern European cultures. Courtesy of the National Gallery, Norway.

CHAPTER ONE

The Mythological Origins
of the Magic Mead of Poetry

*Northern mythology is grand and tragical. Its principal theme is
the perpetual struggle of the beneficent forces of Nature against the
injurious, and hence it is not graceful and idyllic in character, like
the religion of the sunny South, where the people could bask in perpetual sunshine, and the fruits of the earth grew ready to their hand.*

—Myths of the Norsemen from the Eddas and Sagas[1]

The Norse religion was already ancient in medieval times,
though it didn't necessarily fit our modern definition of
religion. Its mythology was ever-changing, and was passed along orally
from generation to generation, primarily by skalds (poets and story-
tellers heavily influenced by psychotropic mead and ale), to provide
entertainment and share knowledge during the long, dark, and cold
winter nights and harsh, lengthy sea journeys. Yes, some elements
remained essentially the same, but these stories weren't set in stone.
Each telling brought something new.

Scribes eventually documented these stories, but often they were
translating ancient writings in Christian monasteries, so the temptation
for embellishment to align these "heathen" myths with Christian belief
was always there. It's not possible to know for sure how much of Norse
mythology was Christianized in translation. While some transcribers
may have been interested in preserving the historical accuracy of what

they were transcribing, many texts have very clear Christian themes imposed upon them. Sometimes it was indicative of a blatant goal to rewrite pagan history into something more Christian; other times it was more subtle. There are many versions of Norse mythological stories available to the modern reader—each with their own nuances. Since no direct written versions of the myths exist, the stories contained within are often contradictory, containing actions and characters that don't appear in other versions. I have compiled the stories I present here from various adaptations. Although I am committed to being as accurate as I am able, my goal is to tell a compelling story while keeping with the overarching theme and main characters. The myths changed over time as they were passed down orally, and based on the narrative technique of the storyteller, so I see no reason not to continue that tradition.

In order to learn how to make mead like a Viking, it is necessary that we discuss the integral role mead played in Norse mythology. It may help to return to the exercise in the introduction to ensure you are focusing on the mind-set of the time period in which these stories were told. Whether or not you choose to do this, consider that none of the "scientific" advances of medieval times—much less Renaissance, industrial revolution, or modern times—were even remotely considerations of ancient peoples. Their impetus was to protect their family (which extended to their village or warband) by ensuring food, lodging, and defense for their homestead. With minimal knowledge of the world surrounding them, and nothing akin to the modern scientific research we are bombarded with, they relied on their senses and the teachings of shamans, sages, and herbalists to assess how the world worked, how they should react to events around them, and how they should nurture and heal themselves both physically and spiritually.

This was how religion and myth came about—a desire to understand the world around us. In essence it was proto-science. Is using gods, elves, trolls, dwarfs, and other mythical beings to explain unseen forces really all that different from our modern reliance on scientists' descriptions of atoms, molecules, bacteria, and other things that can't be seen with the human eye? To modern sensibilities, yes—but remember, we're looking at things from a completely different perspective. When early humans saw natural forces at work, they wanted to understand how they occurred, just as we do. Instead of relying on the writings of scientists, they put their faith in shamans who underwent rituals that allowed them to delve deep into the human psyche via meditations and spirit quests heavily influenced by psychotropic plants and mushrooms. As alcohol is a natural preservative, many of their tonics and potions were created and maintained via ongoing fermentation rituals. Fermentation is the oldest method for preserving food and drink, with the added bonus that it increases the potency of whatever naturally occurring narcotic, psychotropic, or healing effect is already present in what is being fermented.

The Symbolic Structure of the Norse Mythological Worldview

Norse mythology is an epic on the grandest scale, and mead permeates much of its canon. Its personae have served as inspiration for fantastical stories from when they were first put down in writing (after generations of being passed down through oral storytelling) to modern fantasy books and movies.

The mythological structure of the Norse worldview is represented by a great tree, Yggdrasil (*IG-druh-sill*). This tree's roots stem from the Well of Urd, or the Well of Destiny (*Urd* is the root of the Old English *wyrd*, or "destiny"). In the branches of the Yggdrasil tree is Asgard (garden of the sky gods), where wise, conniving, warring—and often very human—gods reside. Jotunheim is the home of giants, who epitomize the impersonal and destructive primordial forces of nature, and also represent evil that can never be fully quelled. These two realms lie above Midgard, where humans reside. Midgard translates roughly to "the middle garden."

It makes sense that Asgard and Midgard, the two most important realms of Norse mythology and those that are most beneficial to humanity's well-being, share the root word *gard*, which is also the root for our modern *garden*. The production and preservation of food and drink were vital to survival in the harsh climate of Northern Europe, with its long, dark winters. Locating and selecting an area in which to settle and build a homestead required careful thought and planning, and could result in disaster if the choice was a bad one. Once they cordoned off an area by

VÖLUSPÁ

Hear my words, you holy gods,
great men and humble sons of Heimdall;
by Odin's will, I'll speak the ancient lore,
the oldest of all that I remember.

I remember giants of ages past,
those who called me one of their kin;
I know how nine roots form nine worlds
under the earth where the Ash Tree rises.

Nothing was there when time began,
neither sands nor seas nor cooling waves.
Earth was not yet, nor the high heavens,
but a gaping emptiness nowhere green.

Then Bur's sons lifted up the land
and made Midgard, men's fair dwelling;
the sun shone out of the south,
and bright grass grew from the ground of stone.

. . . .

fencing or other means, this was their home. Anything outside was subject to chance—or, as they saw it, the whims of gods, giants, and other
mythical beings such as the Norns, three divine females who reside in
the Well of Urd and determine the destiny of mankind by carving runes
into the trunk of Yggdrasil (or in some versions, weaving mankind's fate
into a tapestry).

Ljossalfheim, more commonly known as Alfheim or Alfheimr (Norse
for "elf home"), is the realm of the light elves, or air spirits—similar to
the faeries of Celtic and Germanic fairy tales. There are other realms,

There is an ash tree—its name is Yggdrasil—
a tall tree watered from a cloudy well.
Dew falls from its boughs down into the valleys;
ever green it stands beside the Norns' spring.

Much wisdom have the three maidens
who come from the waters close to that tree;
they established laws, decided the lives
men were to lead, marked out their fates.

. . . .

She sat alone outside; the old one came,
anxious, from Valhalla, and looked into her eyes.
Why have you come here? What would you ask me?
I know everything—where you left your eye,
Odin, in the water of Mimir's well.
Every morning Mimir drinks mead
from Warfather's tribute. Seek you wisdom still?

—from *Völuspá*, the first poem of the Poetic Edda,
translated by Patricia Terry[2]

which are outlined in the illustration on page 7, but the last one that we need to concern ourselves with for now is Svartalfheim—the realm of black elves, stone spirits, or—as they are more commonly known today—dwarfs. In Norse mythology, dwarfs possessed incredible strength and intelligence and were master crafters of stone and gems (they made many of the gods' most prized magical possessions, including Thor's hammer, Mjolnir). The dwarfs—who were also the source of legends of trolls, gnomes, and kobolds—were malignant, crafty creatures who lived in caverns deep underground.

Yggdrasil (the world tree) symbolizes, among other things, all realms within the Norse mythological worldview. Its concept is much more ancient and universal than the Norse, though. Religions and mythologies from many cultures have a world tree motif that is central to their belief system. *The Ash Yggdrasil* by Friedrich Wilhelm Heine (1845–1921), published in *Asgard and the Gods* by Wilhelm Wagner (London: Swan Sonnenschein, Les Bas & Lowery, 1886), page 27.

How the World Came to Be

Many eons ago, dark mists rose from Niflheim—a region north of the great frozen abyss Ginnungagap and fed by the 12 tributaries of the river Elivagar—and met with the flames of Muspelheim (the land of fire) to the south, resulting in steam that slowly froze into rime and hoarfrost. Over time, the interactions of these opposing elements filled the great yawning gap.

The substance that filled the gap eventually attained consciousness and became the enormous (even by giants' standards) Ymir, which

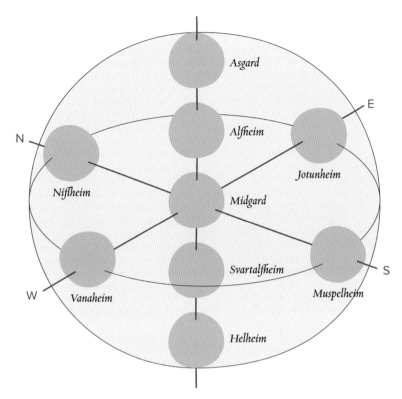

The nine worlds (Old Norse Níu Heimar) of Norse mythology aren't intended to depict a belief in the physical structure of the world. Rather, they are meant to symbolize where each of the spiritual beings of Norse mythology reside. These homes of the gods and other beings can be reached by humans through skaldic meditation (spirit quests), and sometimes manifest themselves in the physical world.

translates to "seething clay," and the great cow Audhumla, the nourisher. Audhumla, true to her name, provided Ymir with nourishment from the milk that flowed from her abundant udders into four rivers. Audhumla also needed nourishment of her own, which she found in salt that had gathered on some nearby rime. As she licked salt from the block, the god Buri (the producer) emerged; he had apparently been biding his time in the ice block, waiting for the tongue of a giant cow to set him free.

While Audhumla was feeding, Ymir fell asleep. The heat continuing to rise from Muspelheim caused him to perspire, resulting in the birth of a son and a daughter from his armpits, while his sweaty feet produced

The conflict between fire and ice (and subsequent chaos upon their meeting) is a recurrent theme in Norse mythology. Given the environment in which these stories came to fruition, this makes perfect sense. Here, the stratovolcano (or composite volcano) Bárðarbunga erupts in 2014 under Vatnajökull, a massive Icelandic glacier, demonstrating how this environment endures even into modern times. Photo courtesy of Peter Hartree, Wikimedia Commons.

the six-headed giant Bergelmir, who went on to father the race of frost giants. The frost giants promptly took a dislike to both Buri and his son Borr and began a war that lasted for generations.

Eventually, Borr married a giantess named Bestla, daughter of the "thorn of evil" Bolthorn, an alliance that turned the tide. Borr and Bestla bore* three sons: Odin (spirit), Vili (will), and Ve (holy). Borr and Bestla's sons were gods of great power, and combined forces to slay Ymir. The wounds they inflicted on the behemoth triggered a gush of blood that drenched the land with massive flooding. The floods wiped out the race of giants except for two—Bergelmir and his wife—who escaped to the mountainous wilderness of Jotunheim and went on to produce more giants.

---

* Interestingly, our word *born* comes from Borr (sometimes spelled Bor).

After slaying Ymir, Odin and the other Aesir (brash, reactionary, law-abiding, patriarchal gods and goddesses) fashioned the physical world of Midgard from his corpse, recruiting four dwarfs to hold up its four corners: Austri (East), Vestri (West), Nordri (North), and Sudri (South). Over time, the Aesir clashed with the elder Vanir gods (peace-loving, slow-moving, freethinking, matriarchal, fertility-loving gods).

How Mead Came to Be

After a long war in which neither side gained any real traction, the two tribes of gods decided to reconcile. They chewed up some berries, spat into a cauldron (the communal mixing of saliva is an ancient reconciliation tradition), and let the berries ferment. But rather than the intended fermented beverage, what rose from the cauldron was Kvasir, the wisest of all beings.[3] It is unclear whether Kvasir was a man, a god, or another being altogether, but the poems he spoke and the stories he told made even the hardest-hearted of gods and men weep. His words were like pure gold and caused his listeners to transcend into previously unknown realms of the mind; he could wisely answer any question put to him.

The dwarfs, being dwarfs, saw Kvasir as a great treasure and coveted him for themselves. Two particularly malicious dwarfs, Galar and Fialar, kidnapped Kvasir; once they had him in their deep caverns, they murdered him and drained his blood into three cauldrons, later telling the gods that Kvasir had "suffocated from an excess of wisdom."[4] Now that the dwarfs possessed the great wisdom of Kvasir, they sought to preserve it, not because they wanted to gain and share wisdom, but because they wanted to hoard it for themselves.

Knowing that when mixed with the proper amount of liquid, honey would transform that liquid into a magical, ecstasy-inducing substance, the dwarfs added honey to the cauldrons containing Kvasir's blood. Thus was born ancient, mystical mead, which could turn anyone who drank it into a poet full of charm and with a magnificent singing voice—you know, kind of how you think of yourself when you sing karaoke after a few too many drinks. Of course the dwarfs had no intention of sharing it, nor did they taste it.

The three vessels were named Othrorir (inspiration), Son (reparation), and Boden (offering). Othrorir, or "mead of poetry," was what mead came to be called over time (it can also refer to the vessel in which mead is held), as it inspired wise words through poetry and song by skalds skilled enough to know how to pass along the wisdom gained from drinking it. These dwarfs, though, were no skalds. Emboldened by their success, several of the dwarfs traveled throughout Midgard, pestering any humans they met along the way, and then sought to enter Jotunheim to pick on the giants.

Their first victim was Gilling, a simple-minded giant whom the dwarfs came across sleeping on a steep bank, and rolled into the water to drown. Not content with the level of wickedness to which they had already stooped, they gleefully went back to Gilling's house, screamed that he was dead, and dropped a millstone onto the head of his distraught wife as she came running out weeping for her husband's death. The nasty little buggers were on a roll, and continued to wreak havoc throughout Jotunheim, raucously singing songs about their great cunning and the stupidity of the giants (maybe they took a sip of the Magic Mead after all).

Not all the giants were as stupid as the dwarfs assumed in their arrogance, though. One particularly pissed-off and crafty giant, Suttung—Gilling's brother—finally had enough. While the dwarfs were engaged in one of their drunken songs, Suttung snuck up and captured them. He then took them out to a rock in the sea during low tide and set them on it. As the waves began to rise to the terrified dwarfs' necks, they tried anything they could to save their hides. They offered gold, jewels, and other items of beauty—but giants have no interest in flashy bling. As the water continued to rise, the desperate dwarfs offered the Magic Mead. Realizing this was a coveted item the giants could use as collateral in their war against the gods, Suttung agreed to take the mead, tossed the dwarfs in the boat, and rowed back to land. He then brought them back to Svartalfheim and bellowed into the caverns his intent to squash the heads of his hostages between his fingertips if the cauldrons of Magic Mead weren't brought to him immediately. He then plopped down to wait, tormenting the dwarfs so that their screams would make it clear he was serious.

Realizing that not submitting to Suttung's requests would pit them as adversaries of both the giants and the gods, the dwarfs decided they were finished with this game and brought the jars of Magic

GIANT SUTTUNG AND THE DWARFS.

Suttung places the dwarfs on a shoal, threatening to let them drown if they don't give back the Magic Mead. *[The] Giant Suttung and the Dwarfs* by Louis Huard. Published in *The Heroes of Asgard: Tales from Scandinavian Mythology* by A. Keary and E. Keary (London: MacMillan & Co., 1900).

Mead to the surface. Upon bringing the mead to his Great Hall—deep within the caverns of a mountain—Suttung placed a spell on his daughter, the beautiful giant-maiden Gunnlod,* causing her to appear as a hideous witch with long teeth and sharp nails, and locked her in a small

---

* The name is Old Norse for "invitation to battle."

cavern with the Magic Mead, with a directive to "guard it night and day, and allow neither gods nor mortals to have so much as a taste."[5]

How Odin Stole the Magic Mead

Odin—a complex god, and the eldest and wisest of the Aesir—was part trickster, part shaman, part warrior, and part poet. Due to his skills in animism, he could take many forms. At times, he would travel Midgard disguised as Vegtam the Wanderer (a direct influence for Tolkien's Gandalf), an old man with a dark-blue cloak, a long white beard, a staff, and a wide-brimmed hat (part of the brim was kept turned down to hide the socket of the eye he sacrificed at the well of Mimir to gain great wisdom).

Vegtam was often accompanied by two ravens—Hugin and Munin (from the Old Norse *Hugínn*, for "thought," and *Munínn*, for "desire")—who traveled between Midgard and Asgard as his messengers and spies. When his ravens brought to him word of the misdeeds of the dwarfs and how Suttung had thwarted them, Odin immediately began plotting with the other Aesir to acquire the Magic Mead and bring it back to Asgard. As the other gods prepared vessels to hold the mead, Odin closed up the caverns of the dwarfs so that they would never again be able to enter the world of men, and went forth disguised as Vegtam the Wanderer to infiltrate the Hall of Suttung (now known to the gods as "the mead wolf").

This was no simple task, as Suttung, son of the fire giant Surtur (who, in time, will battle with the gods at Ragnarok and bring about the end of this world so that a new one may arise), resided deep in a mountain cavern that was closed off to the outside world save for one entrance, which was guarded by a formidable dwarf sentinel. As the first step in his carefully crafted plan, Odin approached nine thralls (literally, "an unfree servant") belonging to Baugi, Suttung's brother, who were swinging their scythes in a field to little avail. Upon seeing Odin/Vegtam, the thralls asked him if he would go tell Baugi that their scythes were dull and that they would stop mowing until they were provided with a whetstone. Conveniently, the mysterious wandering old man had a whetstone on him. Upon whetting their scythes, the grass practically fell down on its own with barely a swing. The thralls then begged Odin to give them the whetstone, which he did—by tossing it over a wall. The thralls all jumped over the wall in pursuit of the whetstone, several of them wounding one another with their sharpened scythes. Names were called, blame was passed, and they fought among themselves, all dead or mortally wounded in the end.

Odin, who appeared as a giant to giants and a man to men, then recovered his whetstone and paid a visit to Baugi to ask for supper and lodging. The giant—who followed the Norse tradition of providing

Odin sits upon his throne accompanied by the ravens Hugin and Munin ("thought" and "desire") and the wolves Geri and Freki ("the greedy one" and "the ravenous one"). As part of the Norse shamanistic belief system, Odin was often accompanied by "familiar spirits" in the forms of animals, and would sometimes leave his physical body to visit other realms in the form of an animal or other being. Illustration by Ludvig Pietsch (1824–1911). Published in *Manual of Mythology: Greek and Roman, Norse, and Old German, Hindoo and Egyptian Mythology* (London: Asher and Co., 1865), plate XXXV.

hospitality to strangers—acquiesced, offered Odin a place to rest, and brought him some food. During supper a messenger arrived, who reported that Baugi's thralls had been found dead. Frantic about the potential loss of hay for the winter, the giant panicked. But since this was all part of the Aesir's plan, Odin offered to work for him.

Baugi was initially reluctant to hire a single person to mow all his winter hay, but the stranger offered to do the work of nine men in a day to prove himself, so he accepted. When Odin proved he could live up to his end of the deal, Baugi implored that he stay for a payment of Odin's choice at the end of the season. Odin returned as winter set in, having harvested all the wheat, and demanded his reward—a draught of the Magic Mead.

Distraught, Baugi first claimed he didn't know where it was, but when Odin reminded him that Suttung possessed it, he begged Odin to ask for another reward, as he feared Suttung greatly. Odin was insistent, so Baugi went to his brother, saying that he was in a bind and needed just a draught of the Magic Mead. Exploding in rage, Suttung called his brother an oaf and a fool and other terrible names and told him he would give none of the Magic Mead to one of the Aesir, as he had quickly deduced that only one of their sworn enemies would have asked for this reward.

Returning to his home, Baugi implored Odin to ask for another reward, as he knew of no way into the mountain that didn't involve battling hordes of angry giants. Odin, however, was already one step ahead. From his cloak, he pulled out an auger, and demanded that Baugi use all his strength to bore a hole into the mountain. Baugi took the auger and started boring, then quickly pulled it out, claiming he had completed his part of the bargain, and prepared to leave. Odin, not easily fooled, blew into the hole, causing bits of dust and rock to fly into his face. He handed the augur back to Baugi, telling him to finish the job and to knock off his poor attempts at trickery. Grumbling, Baugi took the auger, bored some more, and handed it back. Trusting he had done the job this time, Odin changed into a snake and wriggled through the hole, narrowly outracing the auger that Baugi thrust in after him in an attempt to avenge his damaged pride.

Once he arrived in Gunnlod's cave, Odin changed back into his full godly form in an attempt to woo Gunnlod so she would take his hand in marriage. He succeeded, and she lay with him and pleasured him on her couch for three full days. So entranced was she with Odin and his godly powers in the sack that she allowed him to drink his fill of the Magic Mead. He did, finishing off all three vessels. His job complete, he bade Gunnlod farewell, changed into an eagle, and flew back out of the hole.[6]

Halfway to Asgard, Odin realized he was being pursued by Suttung, who had also transformed into an eagle. But fortunately for Odin, the other gods had been planning for his return. When they saw Suttung closing in, they prepared a massive pile of combustible materials just within the walls of Asgard. They set fire to the materials as Suttung flew over, singeing his wings and causing him to fall into the fire and perish. They had also set out three vessels, into which Odin regurgitated the

An image of Suttung and Odin as eagles from the 18th-century Icelandic manuscript "SÁM 66." We see Odin spewing the Magic Mead into three waiting vessels while letting some drop to Midgard from his . . . other end. Courtesy of Árni Magnússon Institute, Iceland.

mead. In his rush to spit it all out, he let three drops fall to Midgard, which were subsequently discovered by humans.

This excess mead was known as the "rhymester's share" and the recipients were "poet-tasters." In addition, at times Odin intentionally gifted mankind with a portion of the mead that had been reserved for the gods. These people went on to become renowned skalds; in gratitude they worshipped Odin as the god of poetry, song, and eloquence. In some variations of the story, Odin didn't swallow the mead, while in others, he swallowed it and regurgitated it. I prefer the latter, as it emulates what bees do: regurgitating nectar to turn it into honey.

CHAPTER TWO

Mead in the Viking Age

May abundance of mead be given to Maelgyn of Anlesey, who supplies us
From his foaming mead horns, with the choicest of pure liquor.
Since his bees collect and do not enjoy,
We have sparkling mead, which is universally praised.

—The Mabinogion[1]

In order to understand how the Vikings made mead and other fermented beverages, it is necessary that we look at the brewing traditions of other early Northern European cultures. An ideal point of reference is the Anglo-Saxon era (approximately 410 to 1066). The reason for this is threefold:

1. The Vikings left little verifiable evidence in regard to their actual brewing techniques.
2. The word *Viking* is really little more than a loosely defined term given to a multinational group of explorers, plunderers, and settlers who traveled the breadth of Northern Europe, briefly settled North America, and journeyed to Normandy and Spain in Southern Europe.
3. As a direct result of the second point, they increased their numbers (primarily during the Anglo-Saxon era) from an initial group of hardy Scandinavians, to additions (willing and unwilling) culled from a hodgepodge of nationalities, including the Celts, Russians, and various Germanic tribes (namely, the Angles/English, Saxons, and Jutes).[2]

Based on the many references to mead we find in Norse mythology, the people we refer to today as the Vikings had established techniques for fermenting alcohol that were passed down through generations long before they ventured in droves to foreign lands. Vikings were curious by nature, though, and would have integrated ingredients and procedural knowledge picked up from travels into their brews. Although there are tantalizing references to the making of mead in the Icelandic sagas and other texts—and archaeologists are learning more about the actual ingredients they used—we can't say for sure *exactly* how the Norse brewed their alcoholic beverages. Fermentation is such a simple process—one that has been practiced with little alteration for millennia—that it's not much of a stretch to take what we know about traditional fermentation practices today, combined with knowledge of ingredients common in ancient brews, and say confidently that we can indeed make mead like a Viking.

Thanks in large part to the cutting-edge research of biomolecular archeologist Patrick McGovern, we know many of the ingredients the Vikings used in their brews, as well as the types of vessels they likely employed to brew and serve alcohol. McGovern and his team at the University of Pennsylvania's Biomolecular Archaeology Laboratory for Cuisine, Fermented Beverages, and Health have conducted in-depth microbiological analysis of fermented beverage residue on serving implements found in the graves of Bronze and Iron Age Nordic peoples (the probable ancestors of the Vikings). From this research, we know that what the Vikings most likely made with frequency was a highly complex "grog" made up of several blended ferments with multiple ingredients. Northern Europeans weren't the only group to have employed this technique. As McGovern notes in *Uncorking the Past* when referencing his analysis of graves in the Neolithic village of Jiahu in China's Henan province, "The beverage makers at Jiahu were skilled enough to make a complex beverage consisting of a grape and hawthorn-fruit wine, honey mead, and rice beer."[3]

McGovern cites the scarcity of sugar as one reason for the prevalence of grogs in ancient Northern Europe: "Various cereals, which had originated from the Near East, could be sprouted and their starches converted into sugar. Apples, cherries, cowberries, cranberries, lingonberries, and even cloudberries in the far north, were additional, albeit somewhat limited, sources of sugar." He further remarks that "The most convincing evidence for a Bronze Age Nordic grog comes from the grave of an eighteen- to twenty-year-old woman who was buried sometime between 1500 and 1300 B.C. in an oak coffin under a tumulus at Egtved, Denmark."[4]

Analysis of a birch-bark container showed "the remains of cowberries (*Vaccinium vitis-idaea*) and cranberries (*V. oxycoccos*), wheat grains, filaments of bog myrtle (*Myrica gale*), pollen from the lime tree, meadowsweet, and white clover (*Trifolium repens*)." The examiner of the container, Bille Gram, "concluded that the Egtved young woman clearly belonged to the upper class and had been presented with a special mixed fermented drink of mead, beer, and fruit." This heady and highly alcoholic beverage was the claim to fame of the Northern European "barbarians," as they were referred to by the high and mighty Romans—who had already initiated some of history's earliest alcohol laws, requiring that wine be diluted with water to diminish the alcoholic effect. They found the fermented beverages of the barbarians (which undoubtedly were partially responsible for fortifying the hordes who eventually sacked Rome) to be putrid, referring to them as "rotted barley water."[5]

The evidence is clear that the early Scandinavians took great pride in their brewing, and made brews that could easily rival the finest of today's wines. Archaeological evidence suggests that what the Norse brewed was more akin to a highly inebriating soup. Norse, Anglo-Saxon, and Celtic burial sites often contained brewing cauldrons, serving vessels, drinking vessels, and long-handled serving strainers. The strainers were used as serving implements to strain out the bits of fruit, barley, seeds, spice, and various other substances that were floating in the brew.[6] One grave of a Celtic prince in Hochdorf, Germany, contained a massive 132-gallon (500-L) cauldron that had been three-quarters full at the time it was placed in the tomb. Nearly all the 350 liters (92 gallons) of liquid in the cauldron was mead, which had been made from honey derived from the pollen of 60 different plants. The prince was provided not only with plenty of mead, but also with "eight bronze-studded drinking horns with gold and bronze fittings and a ninth one of iron, over one meter long and 5.5 liters in capacity."[7]

The historical peoples of the British Isles—in particular Ireland and Wales—also held mead in high regard. Most honey was set

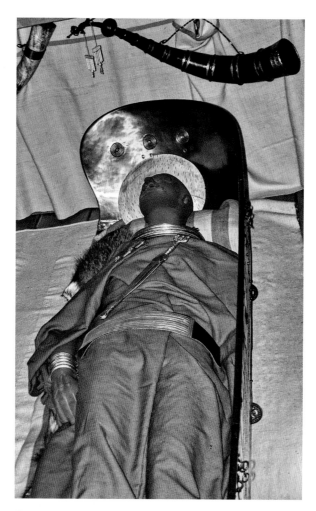

This grave of a Celtic prince in Hochdorf, Germany, contained a massive 500-liter cauldron filled nearly to the brim with mead. Photo courtesy of NobbiP, Wikimedia Commons.

aside for royalty, whose servants used it for cooking, but a large portion was also set aside for the palace mead brewer who, along with the butler, was given special treatment. Laws codified by the esteemed Welsh chieftain Hywel Dda (Hywel the Good) stated, "'The protection of the mead-brewer is from the time he shall begin to prepare the mead vat until he shall cover it,' and that of the butler 'is from the time he shall begin to empty the mead vat until he shall finish.'"[8] The mead brewer and the butler held court privileges, were given housing in the Great Hall, and were allotted a portion of each vat of mead they brewed.[9]

The ancient Celts were so enamored with honey and mead that they named the Hill of Tara's banqueting hall Teach Miodhchuarta (house of mead-circling). This impressive structure, measuring 90 feet wide by 755 feet long, could hold thousands of guests for massive celebrations and important events, which surely involved the consumption of copious amounts of mead. This "Great Hall housed all the leading knights of Ireland, acted as the ceremonial entrance to Tara, and was the point where all the major roads of ancient Ireland converged."[10] The English surely held honey and mead in high regard as well, but not to the same degree as the Welsh and Irish, as there is much less mention of honey in early English law books. They did, though, require that records be kept of hives, bees, and any wild honey discovered in the forest. As far as the royalty and the church were concerned, much more weight was placed on the production and storage of wax for candles to be used in church services and for telling time.[11]

Viking and Anglo-Saxon Drinking Vessels

Although little physical evidence remains of drinking vessels made with organic materials, we have archaeological and written evidence that the Vikings and their Anglo-Saxon counterparts had the tools and technologies for making watertight wood and leather drinking vessels. They also used bone, horn, ivory, antler, ceramic, metal, and glass.[12] The types of vessels made from and decorated with these materials include cups, bottles, flagons (large mugs with a handle, a spout, and sometimes a hinged lid), chalices, bowls, and drinking horns. Another curious vessel was the "clawbeaker." Although German in origin, it was prevalent during the Viking and Anglo-Saxon periods, so the Vikings surely would have drunk from it. "The typical clawbeaker is a conical glass jar with a small foot and an everted rim, to the which are added both a decorative horizontal trail at top and bottom, and a series of hollow 'claws.'"[13]

The drinking horn is what is most often associated with the Vikings. While they drank from other vessels as well, the horn was particularly important to them, and was often used in ceremonies. While most horns probably came from domesticated bovines, those obtained in

This drinking horn buried with a Celtic prince in Hochdorf, Germany measured over 3 feet in length and 5.5 quarts in volume. Photo courtesy of Chez Câsver, Wikimedia Commons.

the wild were a sign of hunting prowess. The most important horns were taken from aurochs—massive beasts from which modern cattle are descended—and were often decorated in gold, silver, and jewels. Extinct since the 17th century, aurochs stood nearly 6 feet high at the shoulders, were highly aggressive, and had curved horns that could grow as long as 2½ feet. "It seems that to be offered alcohol in a horn was a mark of status, although—the many references to drinking horns in the heroic literature apart—clearer evidence comes from later sources including the Middle English romance of *Kin Horn*. At her bridal feast a king's daughter is carrying a ceremonial drinking horn round to the guests, but when she is accosted by a man she thinks is a beggar, she offers him instead drink in a large bowl as being more fitting to his condition."[14] For more on ceremony and status in regard to drinking and drinking vessels, see chapter 9.

What Exactly Am I Drinking Here, Anyway?

In ancient times, there was little distinction between different types of alcohol, which adds to the complexity of interpreting recipes from old texts. Various words were used for alcoholic beverages (as an added bonus, the various spellings of these words have changed over time), and it's not always apparent which modern-day equivalent they are referencing.

As Ann Hagen notes in *A Second Handbook of Anglo-Saxon Food and Drink: Production and Distribution*: "A wide range of drinks, some of which approximated to those we know today, was available to the Anglo-Saxons. The most important, judging by the number of references, were the fermented drinks: *win, meodu, beor* and *ealu*." It gets confusing from here . . . *Beor* was a "cereal-derived drink, being etymologically derived from *bere* (barley)." Sweet honey-based drinks were *ydromellum* or *mulsum*. The most etymologically similar word (and likely closest in ingredients and brewing process) to what we know today as mead was *medo*.

Hagen goes on to explain that "*ealoðo* and *beor* were not synonyms as ale and beer are today, but very different drinks."[15] When we see ale, mead, or beer referenced in the literature of the time, the word chosen to represent strong drink is more likely a literary decision than an accurate representation. "In *Beowulf* . . . the poet refers to *medu, ealu, beor* and *win* . . . In one instance a *medoful* 'cup of mead' is brought out to a drinker . . . 'at the beer drinking' (*Beowulf*, l. 624, 617) and another becomes *meodugal* 'mead-happy' *on beore*."[16]

It can be difficult to classify ancient beverages according to modern definitions, and even today a dizzying array of terms is used to describe mead. While I am fascinated with etymology, it seems that we have an overly complicated glossary when referring to alcoholic beverages that share similar ingredients and fermentation processes. It does help to know what type of flavor to expect when partaking in a vessel of brew—and to have an idea of its contents—but we'll let the brew nerds (in whose hallowed ranks I proudly consider myself to stand) debate the nomenclature. Since we're focusing on mead, I'll use that word primarily when referring to brewing beverages in which honey, water, and yeast are the primary components. As we stumble into other types of fermented drinks on our journey, I'll use the proper verbiage where necessary. When Thorrson returned from a raid, or was looking to pass time during a long, dark, Icelandic winter, he didn't ask his brew-maiden (historically, women did the majority of household brewing—and were well respected for it) for a mug of Riesling. Likely, he said something more like, "Dear wife, could you honor me with a horn of your finest brew?"

Ancient Brewing Ingredients

The list of herbs and other botanicals used traditionally for brewing strong drink in ancient times is long and varied. Many had known medicinal qualities and were used both for that reason and for their preservative and flavoring qualities. These herbs are all still readily available today, either by purchasing, growing in an herb garden, or harvesting wild. Take note that I am not an herbalist and have no medical credentials. My understanding of herbs appropriate for brewing, medicine, and culinary use is based on extensive research, discussions with herbalists and ethnobotanists, and personal experience. Never sample or brew with any plant you haven't identified with 100 percent certainty, check several sources in regard to any contraindications (potential harm if ingested in combination with another drug/herb or an existing health condition), and harvest wild plants sustainably and ethically. Never, ever harvest and ingest a plant that may have been sprayed with pesticides, and avoid harvesting from roadsides.

It is possible that alcohol wasn't the core catalyst for inebriation in all Viking-era booze, at least not for the brews set aside for ceremonial and sacred use (see chapter 9). As noted by Ralph Metzner in *The Well of Remembrance: Rediscovering the Earth Wisdom Myths of Northern Europe*, "In beer and mead, the additives were of much greater importance than the ferment (honey, grains, bread, malt) . . . The pagan 'mead of inspiration' was no simple beer or mead, but must have been a psychoactive beverage whose inebriating ingredients had a stimulating effect upon creativity."[17] Metzner further notes that "The most important component of the magical drink was by no means alcohol, which is really nothing more than a preservative and solvent . . ."[18] It's understandable that Metzner would take this approach, as he is a well-known proponent of the study of transformations of consciousness, and worked with Timothy Leary and Richard Alpert (later Ram Dass) on the Harvard Psilocybin Projects.[19] Patrick McGovern, though, has a more dubious attitude on the subject: "The much-heralded hallucinatory proclivities of Neolithic Europeans are still hypothetical and much in need of confirmatory evidence. As argued by British archeologists Andrew Sherratt and Richard Rudgley, Nordic grog was often spiked with opium . . . , marijuana . . . , henbane, or deadly nightshade . . . The hard evidence for the consumption of such mind-altering drugs in an alcoholic potion remains elusive, however."[20] By no means do I suggest using any of the aforementioned ingredients in brewing; I simply mention them as a historical curiosity.

The following is not an inclusive list, as the plants that can be used for brewing are limited only by edibility, availability, and taste preference. I have listed several that I and other brewers (both modern and ancient) have used with success. Feel free to combine plants for unique flavor combinations and medicinal benefits, with the above warnings about contraindications in mind.

Bog Myrtle (*Myrica gale*, *pors*, or *kynningsris*)

Bog myrtle was used extensively in brewing throughout Europe and was a common bartering item. It was often brewed along with juniper and sometimes with hops, was known to be highly intoxicating, and could produce a hell of a hangover. It provided an exceptionally strong flavor and was described as having a "heady" effect and as "injurious to the health" in a survey of Norwegian brewing traditions.[21] Bog myrtle, though, possesses a wealth of healing qualities, including "expectorant, sedative, fungiastic, and antiseptic properties . . . [it] relaxes bronchial tissue [is] alterative and an effective stomachic [and] can be used as a powder for skin sores and ulceration."[22]

Saint-John's-Wort (*Hypericum perforatum* or *Perikum*)

In a survey of Norwegian farmers conducted by Odd Nordland in *Brewing and Beer Traditions in Norway*, one farmer noted that "*Perikum* was used for spirits, but not for ale."[23] Nordland goes on to note that "Hypericum is used to flavour spirits in many places, all over the country" and also quoted a farmer who stated that "'It happened that someone added a little hypericum, especially if they wanted really intoxicating ale.'"[24] So it does appear that it was used for flavoring ale, as verified by a quote from another farmer: "'In summer, we picked *perikomblomar*, hypericum, and *hardhaus*, yarrow, and then put them in layers with the malt into the filter-vat, now and then adding a juniper twig, until all the malt was in the vat.'"[25] As a final bit of evidence, Nordland notes that the various words for Saint-John's-wort indicate that "the herb is characterized as having a connection with ale, or as a plant to be used like hops."[26]

YARROW (ACHILLEA MILLEFOLIUM)

The sheer number of Scandinavian words and variants for "yarrow" that specifically refer to ale indicates that it was a very popular ingredient for brewing. In addition to *olkall* (ale man), yarrow "has a number of names indicating its use in brewing. This herb is called *jordhumle* throughout Scandinavia; *backhumle*, *åkerhumle*, *skogshumle* also occur; in Denmark, we find *brygger*, *gjedebrygger* [which curiously is pronounced something like *yeti brewer*], 'brewer' and 'goat brewer' in Jutland. Of special interest are Icelandic terms such as *vallhumall* and *jarðhumall*, 'meadow hops' and 'earth hops', showing that brewing with yarrow was practiced there, too."[27] Even today, yarrow is used throughout the world for its medicinal qualities, being "one of the most widely used herbs in the world . . . More than 58 indigenous tribes regularly used it for medicine in North America, and it has been well known throughout Europe since the beginning of recorded history."[28] Yarrow is a diaphoretic herb, and can be used for reducing fever and treating hypertension. It can also be used as an "astringent, anti-inflammatory, and bitter tonic in the gastrointestinal system,"[29] and has been used since ancient times to stanch bleeding.[30] The list goes on.

WORMWOOD (ARTEMISIA ABSINTHIUM)

Possibly one of the most bitter substances known to man, but also one of the most intoxicating, wormwood was a common ingredient in ancient brewing. It was often used for medicinal purposes due to its antibacterial properties, and was common for helping with digestive problems. Cultures throughout time have noted wormwood not only for its health benefits, but also for its sacred properties. Wormwood has long been demonized due to its use in absinthe, which is illegal in its true form in the United States and most of Europe. While it does add to the potency of alcohol, its reputation as a psychoactive or even a poison is overstated.[31]

Its bitterness is actually vital to releasing its effects. In discussing herbs with bitter compounds, herbalist James Green notes: "In certain plants, the bitter principle prepares the way for the other active ingredients . . . Most bitter compounds are soluble in water and, on the whole, soluble in alcohol."[32] I once brewed a simple wormwood ale with honey and malt extract, drawing from a recipe in Stephen Harrod Buhner's *Sacred and Herbal Healing Beers*. Buhner notes that for a 5-gallon (20-L) batch he initially went with ¾ ounce (21 g) of wormwood and that only those with a "tendency toward gustatory sadomasochism, tongue flagellation, or those who enjoy the taste of earwax might find it pleasant."[33] He recommended using only ½ ounce (14 g), so I went with ¼ ounce (7 g). The result was nearly undrinkable even after a couple of months in the bottle. I ended up keeping the bottles for flu season to choke down for medicinal use. A pinch of wormwood early in the boil when making beer (see chapter 8), mixed in with some other herbs is not necessarily a bad thing, though.

MUGWORT (TANSY, *REINFANN*, *TANACETUM VULGARE*)

Mugwort belongs to the same family as wormwood, chamomile, and yarrow, and thus has comparable mild narcotic properties.[34] It also has similar bittering qualities and should therefore be used sparingly. Mugwort, however, is much less bitter than wormwood, and I've used it to pleasing effect in herbal beers. Both mugwort and wormwood were traditionally used to expel worms from the digestive tract. They are therefore also good for alleviating digestive problems, and mugwort can also be used for problems with liver congestion (such as from eating rancid, oily food), as it helps stimulate bile flow.[35] Additionally, by following recipes closely or through experimentation, bitter herbs can complement the sweetness of malt and honey nicely.

ALEHOOF (*GLECHOMA HEDERACEA*)

Most commonly known in the United States (where most consider it a weed) as creeping charlie, this member of the mint family was used as a preservative and bittering agent in beer long before hops became

Alehoof, commonly known throughout the United States as creeping charlie, is a common sight in untreated lawns and wild lands.

the accepted norm. It has many other names, including ground ivy and gill-over-the-ground. After being brought over from Europe, it quickly established itself as an invasive plant and a nuisance to those interested in keeping a well-manicured yard or garden. Alehoof is often confused with henbit (*Lamium amplexicaule*), and sometimes with purple deadnettle. These are also edible plants, and often arrive at about the same time in early spring. Using any of these herbs in brews, teas, and other concoctions—as well as in salads and soups—can significantly improve the immune system due to the powerful antioxidants and vitamins they contain.

PINE AND SPRUCE RESIN

Pine and spruce root chippings are a traditional flavoring agent and fermentation aid for Scandinavian ales and distilled liquors. Historically, resin was also used to keep wine and ale from souring. "The utilization of the resin of the pine and the spruce seems to be very ancient," notes Odd Nordland in *Brewing and Beer Traditions in Norway*. "It is also mentioned in song XX of the old Finnish epic *Kalevala*. When he was brewing for the wedding of Ilmarinen, the hero, the brewer added spruce cones and pine twigs, to start the ale fermenting."[36] Spruce beer was required fare for long sea voyages led by Captain Cook and other sea captains (including the Vikings, of course!) to prevent scurvy, and spruce beer made with molasses fortified American troops during the American Revolution.[37] Several microbreweries today make spruce-tip beers, and you can acquire your own freshly harvested spruce from Colorado (see the resources section at the back of this book).

HOUSEHOLD SPICES AND OTHER FLAVORINGS

Various household spices that were common in ancient Scandinavia and in many homes today can serve as flavoring agents in brewing. The uses are limited only by the brewer's imagination and creativity, and can include spices such as caraway, cardamom, sage, black pepper, red pepper, turmeric, and ginger. Look in your spice cabinet and think of all the possibilities.

DANDELION (*TARAXACUM* SPP.)

See chapter 6 for a discussion on using dandelion petals in flower mead. The stem, leaves, and roots each have their own unique medicinal qualities and can be used in moderation to flavor and boost the healing power of mead. Use them separately, or chop them all up together and add them to the boil when making a honey beer, using the herbal beer section in chapter 8 as a guideline.

Tobacco

I haven't tried it, nor do I recommend it, but I've seen tobacco referenced as an ingredient in mead and ale often enough in my readings that I found it worth noting. Some indigenous South American tribes use tobacco and honey in their sacred rituals. A type of mead is a predominant beverage in the southern regions of South America, and a curious tobacco-honey beverage is common in northern South America, "where, what could be called tobacco 'honey' is made, that is, where tobacco is soaked or boiled and then imbibed in a liquid or syrupy form."[38]

Heather (*Calluna vulgaris*) and Heather Honey

As with hops, heather holds a unique place in the history of mead and ale—in particular the variety common in Scotland and parts of Scandinavia. What is it about heather that made this ale so coveted? Was it the quality of the heather, some secret ingredient we'll never know, or perhaps the heather honey that was likely used in brewing it? Perhaps it was what was *on* the heather that was unique.

The heather found in Scotland is often coated with a naturally occurring moss, referred to colloquially as "the fogg." This "fogg" is noted to

Heather in full bloom near Pitlochry, Scotland. Heather or heather honey from Scotland can be used to make an authentic reconstruction of the heather ale (or mead) of legend. Photo courtesy of Die4Dixie, Wikimedia Commons.

have certain—ahem—psychotropic properties. We can attribute this discovery to Bruce Williams of The Williams Brothers Brewing Co. in Glasgow, Scotland. Williams gathered some heather for the brewery's planned brewing of a heather ale and sent it off to the English botanist and brewing chemist Keith Thomas. The results showed that the white powder was a mild narcotic hallucinogen and contained wild yeast that has been traditionally used to ferment heather ale.[39] I have had the ale in question—which is, sadly, made from washed heather tops. It also, though, contains traditional ancient brewing ingredients such as sweet gale, or bog myrtle. I recall it being quite flavorful.

There is a strong likelihood that heather mead (or ale) was traditionally made from both heather and heather honey. Heather honey is an intoxicating, unique, and highly nutrient-laden honey. When left to sit, it gains a thick, gelatinous consistency, a property that makes it extremely difficult to remove from a hive. Whatever honey wasn't extracted for food (an ambrosia-like food at that) was left in the vat, to which fogg-covered heather tips were added. It would then spontaneously ferment via the wild yeast from the heather and honey. This intense mix of nutrients—from the honey, comb, bees, propolis, heather, and yeast—would have resulted in a mystical beverage indeed.

In his book *Viking Ale*, folklorist and philologist Bo Almqvist presents an in-depth argument on why the mythical heather ale was more likely a product of the Vikings rather than the Picts, to whom it has been historically attributed. The Picts were legendary. A diminutive people, they were fierce warriors and gadflies to the Romans in their attempts to conquer Britain. They had few friends, as they also battled the Celts, Anglo-Saxons, and Vikings. Eventually, as legend goes, it was the Scots who wiped them out—but this is a matter of much debate. Almqvist references a recipe from a Scottish "old coverless book of cottage cookery" that he feels may give some indication as to how heather ale was traditionally made (see the "Heather Ale" sidebar).

As Almqvist notes, "The syrup and the ginger, of course, reveal that this recipe is not particularly old."[40] While this may be the case, it could very easily be drawn from a much older recipe. For instance, perhaps the ginger was a substitute for another pungent root or bittering agent such as yarrow or mugwort? I'll warrant that the syrup may very well have been honey, or perhaps molasses. As of the writing of this book, I have yet to locate heather

Heather Ale

Heather, hops, barm, syrup, water.

Crop the heather when it is in full bloom, enough to fill a large pot. Fill the pot, cover the croppings with water, set to boil, and boil for one hour. Strain into a clean tub. Measure the liquid, and for every dozen bottles add one ounce of ground ginger, half an ounce of hops, and one pound of golden syrup. Set to boil again and boil for twenty minutes. Strain into a clean cask. Let it stand until milk-warm, then add a tea-cupful of good barm. Cover with a coarse cloth and let it stand till next day. Skim carefully and pour the liquid gently into a tub so that the barm may be left at the bottom of the cask. Bottle and cork tightly. The ale will be ready for use in two or three days.

—Quoted in F. Marian McNeill, *The Scots Kitchen*
(London, 1929), p. 237 in the 1961 reprint

or heather mead made from fogg-laden heather, but have made a heather mead from heather I bought at a homebrew-supply store. I have to admit, a glass or two of this mead is highly intoxicating and has influenced some of the more . . . interesting portions of this book. But any edible flower can be used to make a floral mead, which I have outlined in chapter 7.

Hops and Gruit

Rather than discuss hops on its own as an ingredient for mead and honey-based beers, it seems more fitting to discuss its convoluted place in brewing history, particularly in regard to the once-popular beverage gruit (or *grut*).

Up until the Bavarian Purity Law of 1516, or the *Reinheitsgebot*, which required that beer, specifically German beer, contain only water, barley, and hops—yeast was added later when it was discovered that it was a substance of its own—gruit was the alcoholic beverage of the masses.

Gruit rarely contained hops. Gruit was a "combination of three mild to moderately narcotic herbs: sweet gale (*Myrica gale*), also called bog myrtle, yarrow (*Achillea millefolium*), and wild rosemary (*Ledum palustre*), also called marsh rosemary . . . Gruit ale stimulates the mind, creates euphoria, and enhances sexual drive."[41] Additionally, in *The Well of Remembrance*, Ralph Metzner notes that "all the Scandinavian sources on beer and beer brewing list grut beer as the cause for the berserker rage."[42]

It's difficult to pin down exactly what a berserker was, but most sources define it as a special type of Viking warrior who would go into battle either naked, unarmored, or clad in wolf or bearskin, a near-unstoppable machine who would work himself into a frenzy before battle. I would hesitate to say that *all* the Scandinavian sources list gruit as the impetus for causing a Viking to go berserk, but it is certainly likely that a very strong gruit made with "magical" (mind-altering) herbs was what they drank, giving gruit a reputation for causing drinkers to become excitable.

Hops, on the other hand, are a sedative and an anaphrodisiac. Yes, that's right—all those massively hopped beers that have become popular in the late 20th and early 21st centuries actually decrease your sex drive and make you drowsy after a few too many.

So what events took place that caused the majority of the drinking world to go from drinking a brew that made them happy, horny, and ready for battle to one that makes them happy, somewhat horny, and more interested in sleeping than fighting? As with most major changes in the world, it's a convoluted story that involves politics, religion, and laws being made to keep folks from enjoying themselves too much. As Stephen Harrod Buhner notes in *Sacred and Herbal Healing Beers*, "Hops finally gained ascendency in Germany at nearly the same time Martin Luther was excommunicated from the Catholic Church (1520). It is doubtful this is mere coincidence. One of the arguments of the protestants against the Catholic clergy (and indeed, against Catholicism) was their self-indulgence in food, drink, and lavish lifestyle . . . The Protestant reformists were joined by merchants and competing royals to break the financial monopoly of the Church. The result was, ultimately, the end of a many-thousand-years' tradition of herbal beer making in Europe and the narrowing of beer and ale into one limited expression of beer production—that of hopped ales or what we call today beer."[43]

The near-universal adoption of hops as the standard bittering agent and preservative for beer was a slow-moving process. For several centuries, there was significant opposition from the general public and from established gruit breweries regarding the new rules being pushed. For one thing, many breweries had been brewing gruit for years using closely guarded secret recipes, and they weren't about to throw a new ingredient into their recipes that not only increased the bitterness, but also was known for its sedative qualities. The drinking public shared this concern, as they had grown accustomed to the strong, sweet, and slightly sour ale that had been served in public houses for as long as they had been drinking. As noted in *A History of Brewing*, "Traditional ale must have been strong and sweet—the stronger it was, the longer it would keep. Beer, however, was protected by the hop and could be milder."[44] The term *milder* here means that fewer sugars were needed in the mix to break down into alcohol. In addition, greater quantities of water could be used, resulting in a less alcoholic beverage.[45] It's strange to think that, paradoxically, the majority of beer drinkers in the United States today would revolt if laws were put into place that severely limited the production of their cherished watered-down, heavily marketed rice- and corn-based beer in lieu of—let's just say—something with depth and flavor.

Many medieval breweries purchased their gruit ingredients already mixed with the malt from monopolists who had purchased the rights for the most popular recipes, which meant that large, influential economic powers (that is, the church) had a vested interest in repressing the use of hops, while brewers saw it as a way to break free. "In some places, like Cologne, monopolistic rights were associated with gruit, generally the church's. The Archbishop of Cologne possessed the *Grutrecht* [essentially a decree giving certain parties the sole rights to gruit recipes or, in short, a tax] and tried to suppress the use of hops."[46]

In essence, as with laws today in regard to narcotics and alcohol, a combination of religious fervency, political manipulation, and mercantile/big-business interests supplanted a variegated ancient brewing tradition with laws that greatly minimized the ability of the masses to produce and ingest what they desire, all in the name of "public health." Fortunately, the craft-beer resurgence in the United States has resurrected variety in beer following the Dark Ages of the mid-20th century, during which large commercial breweries were given free rein to produce mass amounts of swill that still dominates much of the market today. The prevailing notion is that hops were introduced (and are essential) because of their antiseptic and preservative properties. However, many other ingredients possess strong antibacterial properties and help beer "keep,"[47] including honey.

The Decline and Resurgence of Mead

As alcohol brewing became commercialized in the 18th and 19th centuries, brewing in general decreased greatly as a standard household function and mead began a slow slide into obscurity, being relegated mostly to nostalgic references in stories about olden times. This isn't to say that it wasn't still brewed in smaller quantities, particularly by beekeepers and rural brewers in the British Isles. In America, though, Prohibition and the subsequent rise of large-scale "swill beer" brewing had a devastating effect on mead's popularity. The homebrew and craft-beer movements of the late 20th century have helped to bring it back, though. Many wineries are beginning to produce meads, and meaderies are popping up all over the country.

Mead is back, and with the unrelenting vengeance of a Viking berserker!

This hyperbolic 1846 lithograph by Nathaniel Currier perhaps overstates the very real problems that led to the temperance movement of the 19th century, and eventually to Prohibition in the 20th.

CHAPTER THREE

Honey and the Bees
We Have to Thank for It

Mead is the fermentation of honey, producing a liquor that allows human beings, for a time, to experience sacred states of mind . . . all three things—bees, honey, and mead—confer on humankind some of the immortality of the gods . . .

—*Stephen Harrod Buhner,*
Sacred and Herbal Healing Beers[1]

Without honey we wouldn't have mead, and without bees we wouldn't have honey. Honey has been inexorably linked with mankind's survival in much the same way that fermentation has, and for about as long. It is therefore understandable that it holds just as strong a place in mythology and sacred rituals. Despite our advances in understanding bee behavior, the way of the bee is in many respects still as mysterious and fascinating to humans as it was when they were considered mystical beings and agents of the gods. The more we learn about them, the more we realize these industrious little creatures continue to defy understanding.

A Brief History of Honey Gathering

In her 1937 book *The Sacred Bee in Ancient Times and Folklore*, scholar Hilda M. Ransome noted, "Some of the oldest fossil bees are found in the amber of the Baltic coast . . . . Central Europe was probably the region where the different races of bees developed, for the oldest forms have been found there."[2] Based on her assertion, it isn't much of a stretch to assume that bees were known to Western Europeans from very early times. More recently, we have learned that they may have been around even longer. A fossil was found in northern Burma in 2006 that pre-dated the previous oldest-known reported fossil (discovered in amber from the Cretaceous period in New Jersey) by 35 to 45 million years. While scientists can't say for sure that these fossilized insects were pollinators, they do provide a glimpse into what the ancestors of the modern honeybee were like.[3]

KLOTZBEUTE AND STRAW SKEPS

While primitive honey gathering consisted mostly of foragers taking note of where bees had built hives and climbing trees to reach them (or setting long ladders precariously against a cliff's edge), it wasn't long before a rudimentary form of beekeeping developed. As ancient Euro-peans began to settle down and build homes in the forest, they took note of where bees built nests. Initially, they marked the wild nests as their property to avert other honey foragers from disturbing their find, but it didn't take long for them to begin to invite the bees closer to their homes by hollowing out holes in trees, and hanging hollowed-out logs from the branches of trees near their dwellings. "These hollowed-out tree-trunks, known as *Klotzbeute*, are still used as hives in parts of Ger-many, Poland, Lettland, and Russia."[4] These hives were sometimes left simply as plain logs, but many decorated them, often in the form of the face of an old man with a hat, possibly as a reference to Odin's human form. Today people get very creative with these hives, carving into them elaborate designs representing modern figures such as George W. Bush, Osama Bin Laden, and Marilyn Monroe. Another term for this type of hive, that developed in the eastern United States, is a *gum*, as older gum, or tupelo, trees are naturally hollow due to their tendency to rot from the inside out, and are ideal for this style of beekeeping.[5]

Eventually, people began to craft hives from other substances. Along with the *Klotzbeute*, the straw skep, one of the prevailing forms, is most relevant to our study, as both have Northern European roots (interestingly, the word *skep* is Scandinavian in its etymology).[6] A skep

Klotzbeutes were simply hollowed-out logs with a hole for bees to enter and exit. They were either kept as plain logs or carved into rudimentary (or even elaborate) designs. Photo courtesy of Daniel Feliciano, Wikimedia Commons.

Skeps are one of the earliest forms of man-made beehives, traditionally made from coiled, woven straw. Photo courtesy of Rosser1954, Wikimedia Commons.

is simply a hive made from woven straw. Essentially, it is an overturned basket. Once bees inhabit a skep, they augment the straw's tendency to shed water by layering propolis (plant gums and resins collected by bees) throughout the interior, making it watertight. Skeps were first developed in the Middle Ages and were the most popular form of hive for many years. A key advantage of skeps was their portability—it was easy to move them around to lure in new swarms of bees.[7]

This portability was integral to the introduction of beekeeping to America. German bees—known as dark bees due to their aggressive tendencies—were some of the first bees to immigrate to America, brought over by the British as early as 1621.[8] Later, the Swedes—whose history is intimately associated with the bee and its gifts—brought over probably more beehives than any other group of colonists.[9]

LANGSTROTH HIVES

It wasn't until 1851 when Lorenzo Langstroth discovered the concept of "bee space" that beekeeping became highly efficient and therefore ripe for commercialization. Langstroth, who subsequently patented the Langstroth hive in 1852, discovered through careful observation of bees in his backyard hives that they required approximately ¼ to ⅜ inch for walking or crawling space. As a result, he built a removable-frame hive with frames aligned in the hive no more than ⅜ inch apart.[10] The frames, full of honey, could be removed with little disturbance to the colony and replaced with new frames. This system also allowed bee-keepers to observe the state of their bee colonies, which over time led to the development of treatments for controlling disease and increasing production—and has largely moved (at least in regard to commercial beekeeping) far from the original natural method of beekeeping that Langstroth developed.

Prior to the invention of the modern removable-frame hive, many beekeepers killed off most of their colonies to extract honey at the end of the season, generally through the use of sulfur fumes.[11] Another method referenced throughout much of the literature on the history of beekeeping was to drop the hive—bees and all—into boiling water. This technique is referenced in various medieval sources, but is by no means the most sensible way to extract honey. There are hints in historical documents that beekeepers may very well have discovered

A frame from a Langstroth hive. Photo courtesy of Luc Viatour/www.lucnix.be, Wikimedia Commons.

ways to extract honey without killing off all their bees, but history by its nature is not always reliable. A fortunate by-product of this senseless slaughter was mead, usually made by cooling and then fermenting the boiled water (comb, dead bees, and all) after pulling out most of the combs and extracting their honey.

Although I don't yet keep my own bees, I had a great-uncle who kept Langstroth hives on his farm in Ohio, and have acquainted myself with several beekeepers while planning my own hives. As my tendencies are toward using natural, nonchemical methods in everything I do, I am primarily interested in foundationless Langstroth and top-bar hives. In talking to beekeepers and reading beekeeping literature, I have found that there is much debate as to the effectiveness of these methods. However, the beekeepers I know (and have read about) who use these hives have seen great success with little to no need for chemicals or other unnatural interference. The heart of the matter is that keeping bees alive and healthy is a tough business.

As much as a beekeeper may want to keep bees—and produce enough honey for it to be worthwhile—bees today are a very different species from what our ancestors worked with. This primarily comes down to genetics, viruses, and mites, but there are far too many factors (and theories) to consider than we have room to discuss here. People are losing bees regardless of the methods they employ. Many beekeepers who are going natural are experiencing rates of loss similar to those who treat their hives. In talking to beekeepers and looking through the statistics presented by the Bee Informed Partnership's National Management Surveys, there is no indication that treating hives is any more effective than going natural insofar as preventing colony loss.[12] If treating hives clearly isn't stopping bee loss, then perhaps it is best that beekeepers seek out other avenues for solving the problem.

Beekeeper Michael Bush is a proponent of "lazy" beekeeping: methods that require minimal upkeep once they're up and running. This is not only good for the beekeeper but good for the bees. According to Bush, a bee colony is a complex ecology comprising all manner of yeasts, fungi, bacteria, and insects: "The beehive is a web of micro and macro life. There are more than 170 kinds of benign or beneficial mites, as many or more kinds of insects, [and] 8,000 or more benign or beneficial microorganisms that have been identified so far."[13] Essentially, a bee colony is a "super organism" that, in nature, works as a balanced ecosystem. Treating a hive may kill off some of the "bad" bacteria and mites, but it also disrupts this ecosystem, permeates the wax and pollen, weakens the bees, and creates "super mites." It then becomes necessary to introduce antibiotics to "strengthen" the bees, leading to a vicious cycle. Some natural beekeepers utilize "soft" treatments such as essential oils or food-grade mineral oil (FGMO), or "hard" treatments such as formic and oxalic acid to treat varroa mites, but many are successfully employing purely natural methods with Langstroth hives. Bush feels, as he told me in an e-mail discussion, that "treating is just a result of the human need to 'do something even if it's wrong' and the fear of loss causing irrational behavior."

One aspect of Bush's "lazy" treatment is his use of foundationless Langstroth hives. A foundationless Langstroth has the benefit of uniform frame size, which allows for consistency and minimizes effort when working with multiple hives, while still keeping with the "natural" philosophy. For Bush's foundationless hives, he avoids using bottom entrances, which are common in most Langstroth hives. This is a controversial practice, as many beekeepers feel a foundation is a necessary component of a hive. However, foundation hives have a few disadvantages: They collect all kinds of toxins over time, whether from chemicals the beekeeper introduces or from insecticides the bees pick up on their travels and drop onto the comb; they require additional expense and intervention by the beekeeper; and they prevent the bees from creating natural cell sizes.[14]

TOP-BAR HIVES

Top-bar hives have been around in some form or another for as long as mankind has been keeping bees. The modern Kenya Top-Bar Hive was developed in Africa based on an ancient Greek hive design (similar to a skep) as an economical method for building simple, sturdy hives.[15] Top-bar hives are essentially a combination of the *Klotzbeute* and skep concepts. Like the *Klotzbeute*, they offer an open space to mimic a hollow log, but the bees are provided some guidance through the use of wood bars that are laid across the top of a simple box construction.

Top-Bar Beekeeping: Organic Practices for Honeybee Health, by Les Crowder and Heather Harrell, is a great resource for those interested in top-bar beekeeping. Crowder worked in commercial beekeeping using the

Beekeeper Matt Wilson displays one of his top-bar hives. Photo courtesy of Matt Wilson.

Langstroth method early in his career, but became disillusioned with the chemicals, antibiotics, corn-sugar- and soy-based feeds, and overall mechanization and monocrop-nectar focus of the industry.[16] As he began researching top-bar beekeeping and developing his own top-bar hives, he was able to move completely away from Langstroth hives, and was successful at it for more than 15 years before he lost nearly all his hives to the varroa mite. According to Harrell, writing on the website For the Love of Bees (www.fortheloveofbees.com), Crowder continued to resist dependence on chemical treatment when this happened and, based on the work of a honeybee researcher, instead began using wood smoke to kill mites. After smoking his hives with juniper and creosote, he was able to completely eliminate his mite problem without the use of chemicals. Following this success, he bred the queens from his now mite-resistant hives and captured feral bees to rebuild his hive stock. He does occasionally still see mites in his hives, but watches them closely and has had no major losses following his implementation of these natural techniques.

Top-bars aren't for everyone, though. Opinion varies as to whether or not they are more economical than Langstroth hives. The compilers of the book *The ABC & XYZ of Bee Culture*, 40th edition, are blunt about it in their entry on top-bar hives: "A bad idea in hive construction that appears repeatedly in the beekeeping press, especially in developing countries, is the recommendation to make and use top bar hives . . . Our observations in developing countries suggest that top bar hives are only slightly less cheaper to build than are regular hives . . ."[17] Despite this claim, top-bar beekeeping has many proponents, and is very popular with backyard beekeepers who wish to produce inexpensive honey naturally. Materials required for top-bar hives are minimal, making them an affordable hive that can be constructed with only rudimentary woodworking skills. Despite the simplicity of their construction, though, they do have a bit of a learning curve. This doesn't mean that beginning beekeepers shouldn't start with them, but it may take a year or two to get the process down. Master Beekeeper Matt Wilson addressed the subject succinctly with me in an online discussion: "Top bar hives are simpler than Langstroth, but not necessarily easier."[18] Wilson keeps several successful top-bar hives and is a proponent of both Michael Bush's and Les Crowder's methods, having learned directly under Les Crowder.

Honey for Health and Healing

The health benefits of honey are innumerable and, combined with the healing power of fermentation into beverages, offer a double whammy of healthful goodness. This wisdom has long been known, but tends to get lost in the modern world of manufactured pharmaceuticals by a culture that has very nearly lost touch with its mead roots. As Hilda Ransome wrote, "Man very early discovered that honey was good for his health, and that a sparkling, fermented drink could be made from it, so it can easily be understood that he came to regard honey as a true 'giver of life,' a substance necessary to existence like water and milk. He held the bee to be a creature of special sanctity connected with those things which seemed to him so mysterious—birth, death, and reincarnation."[19]

The healing powers of honey have been well documented by today's scientific community, but have been known throughout time. Honey was a sacred substance to ancient peoples, and still is to many indigenous cultures that continue to hold on to their ancient connections with the land. The Celtic idea of paradise was one in which there were endless supplies of life-giving honey and mead: "Abundant there are honey and wine / Death and decay thou wilt not see."[20]

Although some of its supposed healing effects bordered on superstition, honey was used to treat ailments as diverse as asthma, psoriasis, arthritis, chicken pox, whooping cough, ulcers, boils, gangrene, hay fever, and sore throats.[21] The last two in particular I can personally attest to. Knowing that honey made from local plants and trees is a proven deterrent to allergies (and that it has a soothing effect on the throat), during allergy season I like to make a strong tea of ginger, lemon, and honey that I sip on throughout the day. Adding a healthy helping of mead or bourbon really helps drive the point home. The modern health care industry is beginning to come around to the idea of honey being a powerful medicine, although not all medical practitioners or researchers are sold on this. According to a 1999 *Bee World* article,

> There is a tendency for some practitioners to dismiss out of hand any suggestion that treatment with honey is worthy of consideration as a remedy in modern medicine. An editorial in *Archives of Internal Medicine* assigned honey

to the category of 'worthless but harmless substances'. Other medical professionals have clearly shown that they are unaware of the research that has demonstrated the rational explanations for the therapeutic effects of honey. Many are not even aware that honey has an anti-bacterial activity beyond the osmotic effect of its sugar content, yet there have been numerous microbiological studies that have shown that in many honeys there are other components present with a much more potent antibacterial effect.[22]

Of course honey, miraculous substance that it is, is no panacea. When fermented, and taken as part of a well-rounded diet, its health benefits can't be denied. Particularly if harvested from a chemical-free hive, honey has an ecology of its own that can work wonders when introduced to the complex web of microbes in the human gut. Making mead that is unpasteurized and wild-fermented just adds to the nutritional kick. Honey generally contains too little water to "spoil" (in other words, ferment) on its own, making it a powerful food preservative. While in this state, it is full of microorganisms, many of them inactive due to the high levels of acidity that make undiluted honey antibacterial.[23] Add just a bit of water and this community of tiny creatures wakes up and gets to work. It's up to the mead maker to harness these microbes to help make a flavorful, nutritious fermented beverage, or—as is the case with many mead makers from medieval times through today—to kill off this community through pasteurization or boiling.

Honey for Mead Making

The sheer number of honey varieties, combined with regional and seasonal availability, can make it difficult to choose a specific type of honey for each batch of mead. Generally, when selecting honey for mead making, use the following criteria (in order of importance):

1. Quality
2. Raw, unfiltered, and unpasteurized
3. Locally produced

4. Varietal
5. Price
6. Convenience

You may have to switch the order around, but always consider the first two before even thinking about the rest.

If honey providers say their honey is local and raw (and are being honest about it), then you can be assured it is high quality and thus appropriate for mead. Honey you buy in the average grocery store, particularly if it says it is "blended," is very likely faux honey. It will produce mead that is plenty drinkable because it has fermentable sugars and tastes like honey, but is often of dubious origin. Anyone who has looked into buying a gallon of honey to use for mead knows that honey is not a cheap product. Because of this, some companies in China and their American distributors have taken advantage of loopholes in the FDA's definition of *honey* to produce a product resembling honey that is better suited to the wallet of the average American consumer. This "honey" is very transparent and clear and won't crystallize like real honey, as it has been filtered and all traces of pollen removed. To enable production of large amounts at low prices, it is then mixed with additives like high-fructose corn syrup and sugar-water. Often it also contains traces of pesticides, antibiotics, and other unpleasant substances.[24]

Many legitimate beekeeping companies have taken to using the TRUE SOURCE HONEY label to pledge that their honey is all-natural. True Source Honey is an initiative started by US honey producers to prevent honey from being devalued due to unethical foreign sourcing. Members place the TRUE SOURCE HONEY label on their honey jars to show that they are committed to producing real honey. More than anything, what this tells us is that the best policy is to get to know your local beekeepers, so that you can always be sure you're using local, trustworthy honey.

But what if you want to make several batches of mead, or plan on making mead during the slow season for honey (autumn through late spring)? Often you can still find local honey during the off season, but price and convenience do play a factor unless you're blessed with lots of time and money to hunt down the highest-quality honey regardless of price. I'll admit, price and convenience are often the determining factors when I purchase my honey for mead, but with some caveats. Since I'm fortunate to have stores near my house that regularly stock local honey at reasonable prices, these stores are where much of the honey for my mead comes from. I know it's local, I trust its quality, and I'm not breaking the bank. But for the same reasons, I'm usually limited to sourwood and various wildflower honeys. This is not a bad thing at all, as these are great honeys for mead. When I travel, I like to look for varieties of locally produced honey that I'm not able to get at home.

Honey Varieties

I can't admit to having used all these varieties for mead, but enterprising mead makers will make mead out of practically any true honey they can get their hands on. Keep in mind that this is only a summary. There are far too many variants to list them all here.

Alfalfa

A light "white" honey, alfalfa is light-amber-colored and lightly flavored. In mead making, it is best used for dry or semi-sweet show meads (meads with only honey, water, and yeast) or delicate flower meads. It is high in glucose and low in nitrogen, so it has plenty of fermentable sugars but may need to be helped along with some extra nutrients. It comes from the pollen of alfalfa, a blue-flowered legume that blooms in the summer throughout most of the western United States.

Avocado

A dark honey made from California avocado blossoms, it is rich, buttery, and nutty. It would make an intriguing mead. California's Heidrun Meadery makes an avocado mead that it describes as "a beer-lover's favorite." A mead made with this could certainly handle some strong flavoring adjuncts and hold up well.

Basswood

Made from the blossoms of basswood, a tree that can be found from southern Canada, to Alabama, to Texas, this light honey has a unique biting flavor. Taste before deciding what type of mead you want to make from it.

BLUEBERRY

True to its name, blueberry honey has a blueberry flavor, but not because blueberry flavoring is added. This honey gets its flavor from the nectar of the flowers of the blueberry bush. It is light amber to amber in color and has a rich, well-balanced flavor. Use in any type of mead for which you want honey to be the primary flavoring agent.

BUCKWHEAT

Buckwheat is an early-spring honey, is dark and thick, and possesses a robust malty flavor akin to molasses. Be sure to taste it before making mead from it. Its strong flavor can be off-putting in mead, but can be quite nice if balanced properly.

CLOVER

The workhorse of honey, clover honey is one of the most commonly found honeys. Although it can be made from single varieties of clover, it is generally a blend of several. It is high in moisture content and therefore ferments fairly quickly. It can be used for practically any style of mead.

HEATHER

A thick, amber-colored, high-protein honey, heather honey has been lauded through the centuries in Europe as a powerful medicinal honey. Many legends are associated with intoxicating mead and ale made from heather honey, particularly the honey produced from heather in the Scottish Highlands.

ORANGE BLOSSOM

Although it can come solely from orange trees, orange-blossom honey more often includes pollen from other citrus sources, including grapefruit, lime, and lemon trees. Produced in areas where citrus grows well—primarily Florida, Texas, Arizona, and California—orange blossom is light amber in color with a unique flavor and is therefore best suited for light traditional or floral meads.

SAGE

As there are various species of the sage plant, the flavor of this honey can vary somewhat, but all tend to be mild, well balanced, and sweet. Therefore, it is another candidate for almost any manner of mead.

SOURWOOD

My primary honey for mead making, because it is commonly found in my region of the United States (eastern), but also because it makes a fine mead. Although a bit acidic and astringent in flavor, the honey isn't sour. Rather, it is sweet and well balanced, and has a big aroma. The sourwood tree (also known as the sorrel tree, or lily-of-the-valley) gets its name because of its large, broad leaves, which have a sour flavor when chewed. It is found throughout the Appalachians, and produces one of the most common (and sought-after) honey varieties in North Carolina.

TULIP POPLAR

True to its name, tulip poplar honey comes from the tulip poplar tree, which grows up and down the East Coast. Although a dark honey, its flavor is surprisingly mild. A fine honey for most meads.

TUPELO

Another common honey in my neck of the woods, and another good honey for all kinds of mead making. Tupelo honey comes from blossoms of the tupelo tree, is mild-flavored and pleasant tasting with a lasting aroma, and makes for an equally lovely mead.

WILDFLOWER

Unless beekeepers work to ensure the variety of flower honey they produce comes from a single source (with small traces of other floral compounds), their wildflower honey is likely to be multifloral or poly-floral. Because of this, wildflower honey can vary greatly depending on the season. Look for spring-wildflower honeys, summer-wildflower honeys, and autumn-wildflower honeys. Because the flavor will vary each year depending on factors such as level of rainfall and overall weather conditions, it is difficult to narrow wildflower honeys down by flavor profile. Generally, it is one of the easier honeys to come by and can be used to make a variety of meads.

To Keep or Not to Keep

If at all possible, I suggest you keep bees and make mead from your own honey when you can. Clearly, this isn't possible for everyone, but the more beekeepers there are using natural practices, the better chances we'll have of surviving as a species. Even if you keep hives but don't have the time or inclination to extract much honey from them, you are providing bees a home and good, quality food so that they can continue doing their job—pollinating the plants that produce the air we breathe. Mead making, if approached right, can be part of a holistic approach to playing your part in healing the Earth. All clichés aside, the planet needs our help.

If you don't—or can't—keep bees, I recommend seeking out honey from natural hives. By making mead from real, local honey, we're helping small- to midscale beekeepers make a living, who are in turn helping to keep bees alive and happy (and the cycle continues). When you experience joy from a horn of well-brewed mead, you are also imbibing a paean to the bee, every mead maker's friend.

A good way to source local, naturally produced honey is to get to know some local beekeepers. If they keep bees on a small scale, they may not always have honey available, but bottles of mead can be a great bargaining chip. Beekeepers keep busy, but they're passionate about their craft and will often be happy to show you their hives and talk with you about bees and honey. Schedule a meeting and bring along a bottle of mead. Health food stores that sell local honey by the gallon are another option. If you go in with some friends and offer to buy several gallons at once, you'll find that many stores will be happy to provide a significant discount. Nothing perks up a health food store clerk more than a shopping cart full of honey . . . except for perhaps that herbal tonic under the counter.

CHAPTER FOUR

Preparing for Battle:
Basic Equipment, Ingredients,
and Planning Strategies

Winemaking is a natural, self-fulfilling, self-purifying process . . . It will try to cast off any foul disease that afflicts it, given a chance, given oxygen enough . . . A winemaking book I bought recently says one rotten blackberry will ruin an entire batch of blackberry wine. Not in my experience, not my blackberries. One vinegar fly can turn a cask of wine to vinegar, we read in another. No, no. My flies are not that influential. All utensils must be chemically disinfected and washed in detergents, we read. Well, sirs, I hope you'll rinse them well.

—*John Ehle,* The Cheeses and Wines of
England and France[1]

Tere really are no "rules" to homebrewing, and don't let anyone tell you otherwise. You should of course start with a basic recipe and technique to get a feel for how things work, but in the end brewing is simply a matter of mixing the appropriate ingredients, coaxing in wild yeasts (or adding yeast), fermenting, aging, and drinking. It can be infinitely more complex than this, but it doesn't need to be. If, like me, you want to brew for the pure fun and experimentation of it— and to emulate the way the ancients brewed—you need not purchase all the expensive equipment homebrewing stores, websites, and brew

A sampling of open- and closed-fermentation vessel styles and sizes, along with some of the equipment needed for mead making.

snobs insist you need. Of course you can brew excellent beer, mead, and wine with this equipment, but I would personally rather brew like a Viking than a chemist.

My brewing setup is more akin to what you'd find in a medieval farmhouse than in a chemistry lab. I appreciate (and sometimes use) modern conveniences that allow for higher levels of quality control than our ancestors had, but I try not to rely too heavily on them. The modern equipment I have, I use mostly because I have gathered bits over the years by bartering with folks who purchased it but never got around to actually using it, or used it for a while until they let life get in the way of their brewing. But people made perfectly good alcoholic beverages for years without much of this equipment, and you can, too.

While I fully support buying a basic beer or wine kit to get started in brewing, with a little research (which I've already done for you!), you can come up with a list of equipment that you truly need, as well as equipment that is nice to have. If you walk into a homebrew store without being prepared, you may walk out with more than you really need (even though anyone who's been bitten by the fermentation bug will tell you that you can never have too much equipment). If you walk in with my list and have read up a bit on how things work, feel free to chat with the staff on what type of brewing you plan on doing. They will inevitably love to talk with you about brewing, but keep in mind that they're surrounded with all kinds of fancy equipment and like to talk about how it's all used. You don't necessarily need all this equipment, but give them a listen and stick to your guns if you feel they're suggesting something you don't need. While I have encountered my share of snobbery (for some reason, specialty hobbies tend to bring out the snob in people) in homebrew stores, they're still among my favorite places to visit. For the most part, they're staffed and frequented by friendly, knowledgeable people who love to talk about their passion. If you get a sense that a nose is being hoisted in your general direction, simply pick out what you need, purchase it, and leave. There are other reasons to make sure you're well prepared when visiting a homebrew store. The person working may have a different style of brewing than you, or may prefer to brew something other than what you're brewing and be unable to offer the advice you need, or they may be unfamiliar with and uninterested in brewing the wild and wooly way.

Another option is to forgo the homebrew stores and catalogs altogether. Create a list of needs and wants, build relationships with other wild fermenters and homebrewing mavericks, and start asking around. Join some Internet groups that are populated by folks with the proper mind-set for Yeti Brewing (I've recommended a few at the end of this book), and start browsing and posting your equipment needs on online trade and barter groups. You will inevitably come across folks who have procured brewing equipment and either stopped brewing or never got around to it. Likely, they (or their significant others) just want to get rid of the stuff. Be frugal and be a Viking. Reuse, recycle, and plunder—but be sure to ask first.

Essential Equipment

The equipment you need for making mead varies greatly depending on your needs, resources, and time. What you *really* need to get started with mead boils down to these essentials:

- A glass, ceramic, or food-grade plastic fermentation vessel.
- A stir stick of the appropriate length and diameter to reach the bottom of the vessel.
- A length of cheesecloth or other porous cloth (for open fermentation only).

As a modern person accustomed to drinking mead, wine, and beer that has been aged to some extent, you will likely want to age your

The essential equipment for creating 1- to 3-gallon (4- to 12-L) batches of mead.

mead to bring out subtle flavors and tame any initial sweetness or harshness. I encourage you to do this, although you can make perfectly flavorful mead that has never been bottled. There are a number of options for procuring or making your own long-term aging materials. The basics you will need are:

- A carboy (a 1- to 5-gallon [4- to 20-L] glass or plastic container with a narrow neck to minimize contact with outer air for long-term aging).
- A siphoning tube.
- Funnels of varying sizes.
- A funnel screen, strainer, or nylon mesh bag for filtering sediment.
- An airlock, purchased or homemade (to place in the opening of your carboy), or alternatives such as a drilled stopper with a siphoning tube inserted into the hole, or simply a balloon or condom placed over the opening. Whatever you use should fit snugly in (or over) the opening of the carboy to keep out external air that can cause souring.

The essential equipment for making 5-gallon (20-L) batches of mead.

- Bottles, jugs, or other suitable drinking/storage vessels.
- A turkey baster, wine thief (essentially an oversized turkey baster that is long enough to reach the bottom of a carboy), or needle-less syringe for drawing small test samples from a carboy or jug.

EQUIPMENT CLEANING AND SANITIZING

Modern brewing calls for a great deal of sanitization. While it's important to keep your equipment clean, the Norse and other ancient cultures didn't have laboratory-produced chemicals and wouldn't have tainted their mystical brews with them anyway. To be fair, brewers and winemakers as far back as the early Romans did use burning sulfur as a sanitization agent,[2] but using clean water and reducing contact with air were still the primary methods for keeping wine or mead from turning to vinegar in a bacteria-laden environment. In brewing wild mead, don't fret too much over sanitization, but make sure all your equipment is well cleaned.

I have talked to several brewers who strive to make mead, beer, and wine naturally, and have found that they have varying opinions on sanitizing. Some avoid sanitization altogether while others employ some level of it, but not with the "kill everything that moves" mentality advocated in most brewing manuals. I fall in the middle. I have come to realize that an overly strong focus on sanitization can kill many of the tiny organisms that you *want* to be part of your brew. If you wish to brew a flavorful and nutritious batch of mead, consider that you are working with an ecosystem of many tiny creatures that are here to help you if you treat them right. Provide them with the appropriate environment in which to flourish and play, and they will return the favor.

I've come across stories from wild brewers who go as far as to stir the must with their hands, put in beard hairs for wild yeast (as in Rogue Brewery's "Beard Beer"), or even sit in a vat to impart wild yeast from their bodies. If you're the type who cringes when hearing this sort of thing, consider first that this only sounds unclean by modern standards. From wine grapes traditionally being crushed in a vat by feet, to Belgian breweries aging beer barrels—which are sometimes left open during part of the process—in dusty, musty, and cobweb-infested cellars, this is a time-honored technique.

Fermentation itself is a proven process for removing toxins, so mead or beer produced in this manner is by no means hazardous to your

Sanitization Options for the Modern Brewer

One Step No Rinse Cleaner

The least environmentally intrusive of the sanitizers, although it is technically a cleanser, and works by cleaning with oxygen. It contains sodium percarbonate, an adduct (two combined molecules) of sodium carbonate and hydrogen peroxide. Once air-dried it is nontoxic and requires no rinsing. The label on the container does have warnings about avoiding contact with eyes and not swallowing, so use with caution. Hydrogen peroxide can be used as an alternative. Use by adding 1 tablespoon (15 mL) to 1 gallon (4 L) of warm water, shaking well, and pouring over equipment. You can also submerge smaller materials in it. Although it does its job in two minutes of contact time, it doesn't hurt to keep equipment and materials in contact with it for longer.

Five Star/Star San

A acid-based chemical used by the food industry for sanitizing surfaces. It works well for killing pretty much anything, including any animals or people who ingest it and don't seek medical help immediately. Many brewers swear by it. Dilute 1 ounce (30 mL) in 5 gallons (20 L) of water and use with caution.

health. One avid mead maker I spoke with, Marissa Percoco, likes to refer to the "micro-herds" of bacteria that we work with as mead makers. She feels we shouldn't think of them as our enemies, but befriend and honor them for what they provide us. To her, a clean, healthy environment is more important than sanitization. Her advice is simple: Clean up after yourself, wash your equipment when needed, and keep your nails trimmed. She doesn't sanitize her fermentation equipment in any manner and has made countless meads, none of which has gone "bad." The one exception she does make is using One Step to clean her bottles. One Step—an oxygen-based, no-rinse, powder-based cleaner that produces hydrogen peroxide when mixed with water—is not

Powdered Brewery Wash (PBW)

Touted as environmentally friendly and biodegradable, PBW is used by many commercial breweries and microbreweries across the country. It was originally developed as an alternative to acid sanitizers, as it is a noncaustic alkali cleaner that won't dissolve copper. You can use it to remove tough stains and deposits by soaking equipment in a solution of it overnight. Dissolve 1 to 2 ounces (30 to 60 mL) in 1 gallon (4 L) of warm water.

Iodophor (BTF and IO Star)

An iodine-based cleaner. This is another no-rinse, environmentally friendly cleaner. Dissolve ½ (BTF) to 1 (IO Star) ounce (15 to 30 mL) per 5 gallons (20 L) of water, allow at least two minutes of contact time, and allow equipment to air-dry.

Household Bleach

An old homebrewing standby and one of the most likely items to find in a typical household. Dissolve 1 teaspoon (5 mL) in 5 gallons (20 L) of warm water, wear safety glasses, and avoid wearing clothing you don't want to be turned into homebrew tie-dye. No matter how hard you try not to, you will likely splash some. It's not the most environmentally friendly substance, though, and equipment sanitized with it must be thoroughly rinsed within a few minutes of adding bleach water.

technically a sanitizer and contains no toxins that will harm your mead or the environment. The reasoning for Marissa's choice is more a matter of time management than anything. Inevitably, empty bottles get stored in places that gather all kinds of dust and gunk. It can be time consuming to scrub them all individually. When I store my bottles in my basement, I always keep them in a covered box or turn them upside down to keep crud from falling in. In preparing to bottle, I take a look inside each bottle and set aside any with residue that won't clean off easily. Unless I'm particularly attached to the bottle and want to take the time to clear it out with a bottling brush, it goes in the recycling bin. Otherwise, I run the bottles through a dishwasher or rinse them with a solution of

½ ounce (15 mL) of One Step to 1 gallon (4 L) of water. Marissa, like many wild-fermentation enthusiasts, is passionate about keeping alive the bacteria that are responsible for our very existence. "The war on bacteria has to stop because without bacteria we would die," she told me, proceeding to gleefully exclaim, "Embrace the micro-herds!"

I'll admit I have accumulated various sanitization chemicals from my early days in brewing beer that are still in my brewing inventory. I don't adhere to the strict "sanitize everything and kill all evil bacteria and wild yeast" mantra most homebrewing books proselytize, but neither do I avoid sanitizing altogether. When starting out in wildcrafted mead making, I slowly weaned myself off the use of chemical sanitizers. I have made several batches using no sanitizers at all and they have turned out just fine. Sometimes I'll sanitize my carboy before racking (transferring fermenting mead, wine, or beer from one vessel to another), or my bottles before bottling, simply because they've spent a lot of time in my dank basement, and because long-term aging is the period during which I envision the chances of infection are greater. More and more, though, I've been going the route of using One Step or iodide or simply rinsing everything out with water.

For my wild-fermentation equipment, I never use sanitizer. My crocks and stir stick I rinse with water or simply wipe down with a clean towel, as I have built up yeast strains on them that I want to keep for future brews. If you want, you can sterilize your equipment. The simplest method is to wash it in hot, soapy water, rinse well, and pour boiling water over everything. Keep in mind, though, that boiling water will kill organisms you may actually want to be part of your brew. I encourage you to go with whichever route you are most comfortable with. A word of warning: If you do use sanitizer, be sure to rinse well. Fermentation can stop dead in its tracks if any bacteria-killing chemicals are present in your fermentation equipment.

Ingredients

The single most important factor in starting out a new batch of mead is the quality of ingredients you use. Use fresh ingredients procured from natural sources, preferably those you have grown or wild-harvested yourself. If you buy ingredients from the store, pay attention to the label and the date. Choose fresh-looking, local ingredients that don't have any chemicals or preservatives. You needn't use products that have the ORGANIC label to make mead naturally, although generally it is best to do so (the word has been corrupted by the corporate world, and many foods advertised as organic hardly resemble what the pioneers of the organic movement meant for it to signify).

While I always encourage mead makers to go ancient and brew local, we do have some modern conveniences that we can use to our benefit, as well as access to ingredients from outside our growing region or season. Personally, I like to focus primarily on natural, wild, and local ingredients that can be found easily in farms, gardens, and wild lands. You know—how the Vikings would have done it. This also enables me to lessen my reliance on the corporate economic infrastructure, and to have more knowledge of the health benefits of each mead I make. That being said, I do occasionally go on a Viking raid and pick up ingredients local to another area. When traveling, always plan on seeking out ingredients to bring home and ferment, or to ferment while on the go. Over time, each mead you make from your travels will remind you of the region from which it was gathered. I'm not talking nostalgia here, I'm talking about real, tangible subtleties in flavor profile, aroma, and personality.

WATER

It may seem like the simplest ingredient in mead, but the type of water you use, the amount you use, the periods of the brewing process during which you add it, and how much you heat it (if at all) will play a significant role in the final product. To put it simply, if it tastes good to drink, it's good for mead.

However, if you happen to enjoy drinking tap water, beware. Tap water has additives that won't necessarily result in bad mead, but may have subtle effects on its overall flavor, and on proper fermentation. Don't fret too much, though. Use clean, potable water and you should be okay. If you're using tap water, run it through a water purifier with a carbon filter or let it sit out overnight to allow any chlorine to evaporate. However, some water systems use chloramine, which is created by mixing chlorine with ammonia, and chloramine won't evaporate as easily as chlorine. Also, ammonia can be detrimental to the flavor of your mead. Always be sure to ask your local water provider what type of chemicals they use if you're going to drink or brew with tap water. Alternatively, let the water come to a boil and then shut the heat off and let it drop to the proper temperature for the type of fermentation method you'll be employing.

Generally, I prefer to go with spring water. If you have access to a spring and have tested its mineral content or determined through your own intuition that it may be lacking in minerals, you may want to add some. If you know that your spring water has low mineral content, you can add a pinch of salt, zinc, calcium, or magnesium for each 5-gallon (20-L) batch of mead. This won't necessarily have a huge effect on your mead, but it also doesn't hurt. Every tiny step you take to give your mead extra care and ensure you're providing an environment for yeasts to thrive is one more step toward an amazing batch of mead.

If you're using good, clean water, there is no need to heat it when making mead, as the average room temperature (60–80° F [15–27° C]) is ideal for fermenting mead. This is the temperature at which yeast tends to be the most active. In many older brewing texts, it is referred to as "blood warm" and can be determined by placing the back of your hand lightly on the must (unfermented mead or wine) or wort (unfermented beer). Warming it, however, will help the honey to dissolve quicker. Also, as you work the honey out of its container, you can start adding warm water and swish it around to liquefy the remaining honey and work as much of it out as possible. When doing a no-heat ferment, I'll often place a jar of honey in a pot of water warmed on the stove before pouring it into my cooking pot or fermentation vessel. If your water reaches the boiling point while heating, give it time to cool down. Some recipes call for pasteurizing the must and skimming off any scum (the official technical term, but a misnomer in my opinion) that rises to the top, but this also removes a lot of the nutrients and flavor enhancers in the honey and kills off any wild yeast. You can do it, but you'll miss out on some complex (and desirable) flavor profiles. The other disadvantage to this method is that you're diminishing the natural antibacterial and healing properties of the honey.

HONEY

Honey is without a doubt the core foundation of mead. Books have been written about it, and you can find more info in the first three chapters on its history, mythology, and use in mead, so I'll be brief here. As a general rule, I like to start with a honey-to-water ratio of 1:4 (for 1 gallon of mead, I use 1 quart of honey to 4 quarts of water [for 4 L of mead, use 1 L of honey to 4 L of water]). This results in a semi-sweet mead. If I want mead that is a bit lighter, I'll increase the proportion of water so that the ratio is 1:5 or even 1:8. Generally for the latter I'm also adding ingredients such as fruit, flower petals, or herbs to provide additional flavoring and sugars. If possible, smell and taste the honey first. Make sure you like the taste when using it for your mead, but don't let that stop you from using a honey you feel could use a better flavor profile. You can always taste the mead throughout the fermentation process and add more honey varietals or other ingredients to adjust for optimal flavoring.

TANNINS, NUTRIENTS, AND ACIDS

The ingredients you use for flavoring will vary greatly based on the type of mead you are making. However, there are a few core ingredients—including raisins, black tea, and even tree bark—that will help bolster your mead and ensure a strong fermentation. You'll want to add these ingredients in small amounts, as they are primarily for providing nutrients, tannins, and acid. In small amounts, they don't have a strong effect on the flavor, but can help the body of the mead immensely, as well as acting as preservatives. As Sandor Katz notes in *The Art of Fermentation*, "In addition to flavors and yeasts, botanical ingredients can contribute acids, tannins, nitrogen, and phytochemical 'growth factors' that stimulate yeast growth."[3] Although you can buy these in chemical form, there are many ingredients from nature that will serve the same purpose. I have outlined in table 4.1 natural ingredients you can use in lieu of the laboratory-produced ones you'll often find in wine and mead kits, but don't feel you need to use them in every batch. Any botanical ingredients you add will likely provide all you need for tannins, nutrients, and acids. Research each ingredient you'll be using in a particular batch of mead for its properties, or trust your instincts—always be sure, though, that you know the properties of what you're brewing with, particularly if you're wild-foraging.

For nearly every batch of mead I make, I add a small handful of organic raisins if nothing else. In addition to being covered in wild yeast, they serve to provide nutrients and tannins. I also sometimes

Table 4.1. Botanicals Containing Acids, Tannins, and Nutrients

Tannins and Acids	Botanicals
Tannin	Oak and cherry bark and leaves (and those of other deciduous trees), grape stems, grape leaves, black tea, nettles, fruits (particularly tart and unripe fruits), berries, nuts, red and black beans, spices, coffee beans, hops, grain husks
Malic Acid	Most fruits (varying degrees of acidity), apples, mayhaws, nectarines, cherries, lychees, bananas, mangoes, peaches, tomatoes, strawberries, grapes, kiwis, watermelon, plums, limes, orange peels, rhubarb, pineapples, carrots
Citric Acid	Lemons, limes, oranges, clementines, tangerines, grapefruit, berries (not blueberries), tomatoes, rhubarb, lettuce, carrots, sourdough bread
Tartaric Acid	Avocados, cherries, grapes, lemons, plums
Ascorbic Acid	Rose hips, hibiscus, citrus fruits, spinach, tomatoes, berries

make a tea (which I cool to room temperature before adding) out of the bark or leaves of deciduous trees such as oak, walnut, cherry, maple, or birch, or black tea (one bag or equivalent amount of loose-leaf tea per gallon/liter). Hops and the husks of barley that have been germinated for brewing beer also provide tannins for bragots and honey beers. Take care with these, however, as too much tannin isn't necessarily a good thing. You can also use grape, blackberry, or raspberry leaves (10 to 15 per gallon [4 L] of mead). Take care to not leave bark or leaves in for too long. Immerse them in the must during the initial fermentation, and remove them within 48 hours to avoid imparting too bitter a flavor. The simplest way of doing this is to stir them in and then scoop them out with a spoon. Or you can wrap them in cheesecloth or a nylon mesh bag, immerse the bag or cheesecloth, and pull it out at the appropriate time.

For recipes that call for a bit of an acidic flavor profile (generally meads with little to no additional flavoring except for honey, or with low-acid vegetables or flowers as ingredients), you have a few options. While many wine and mead recipes call for acid blends, these blends are derived from natural substances that you can add yourself. Acids

MARC WILLIAMS AND PATRICK IRONWOOD
ACIDS AND TANNINS

Marc Williams, a mead maker, ethnobotanist, and the executive director of Plants and Healers International (PHI) based out of Asheville, North Carolina, spent several years adding laboratory-produced acids and tannins to his meads, but over time has stopped doing this. The ingredients he adds to each batch for flavoring provide everything needed to produce highly drinkable meads that he tends to drink within five years of brewing.

Marc has an intimate connection with his meads and the plants he uses in brewing them, which he feels is integral to the quality of a wildcrafted mead. Having sat down with Marc for a "meading," I have to say that I agree. The wildcrafted meads he shared with me (both his own and those of other mead makers) had an effervescent, soul-lifting quality to them. Longtime mead maker Patrick Ironwood, a botany and brewing instructor and an associate of Marc's, has tested many methods for using botanicals as a preservative for meads that he has saved for as long as 20 or more years. He has found that meads to which he added tannin and acid hold (and improve) their flavor much better than those without. Patrick also tends to combine commercial blends with natural ingredients—both for the sake of economy (when brewing larger batches), and because the combination can result in a more powerful preservative effect.

from fruits fall into four categories: citric, tartaric, malic, and ascorbic. Generally speaking, citric acids are all that you will need, but the others can also purportedly help speed up fermentation (see table 4.1). When I determine that a mead needs acid, I usually squeeze the juice of half an orange for a 1-gallon (4-L) batch, or add ¼ to ½ cup (2 to 4 ounces [60 to 120 mL]) of orange juice to taste, multiplying exponentially for larger batches. Generally, though, it is best to wait until further on in the process to determine if acid is needed, usually just before bottling. While you can buy pH test strips or acid titration kits to test the acidity of your mead, the best gauge for the Viking mead maker is good old-fashioned taste. Longtime mead maker Marissa Percoco shared her "awesome acid blend" with me that she developed to replace the standard blend that usually comes in wine kits. She cans or freezes grapes at the end of fall for tannic acids, uses apple slices for malic, any kind of citrus for citric, and oak leaves (or sticks if in the fall) for tannin. She claims that it gives her mead a clean, bright balance that commercial blends (which she equates to imparting a "flat McDonald's effect") don't have.

Take sips of your mead all the way up to bottling—siphon a bit in a small cup, or take a sample with a turkey baster, swirl it around, let it linger, and inspect the mouthfeel to determine if it's lacking in anything. If it seems a bit dull, add some tannin. If it's cloyingly sweet with no bite to it, add some citric acid to the fermentation vessel, using whichever citrus fruit you desire. Start with a fairly small amount, stir well, and taste again. Hold off on adding more if you're still not quite sure. Taste again later in the process and keep adjusting until you're happy. If you taste too much acid, give it time if it only seems to be a minimal amount. If your mouth puckers, it's a good idea to decrease the acidity. The best way to do this is to add more honey. You can either add small amounts of straight honey, stirring and tasting as you go, or mix approximately 2 parts honey to 1 part water. Fill a mead horn or goblet while you're fine-tuning the acid and tannin levels. Savor the horn's gift with each adjustment. When your taste buds tell you to pour another horn, you're good to go. Continue fine-tuning until you like the flavor, but don't overdo it. Even though the final product will taste more refined with aging, once the majority of fermentation has commenced, you'll have a pretty good idea of its final flavor. You can always adjust a bit more when preparing for bottling.

In the end, you want to balance sweetness with dryness and tartness in a manner that suits your palate. If it tastes bad, don't tinker with it too much. Mead has a tendency to go through miraculous changes after being aged for a year or more in a bottle. Above all, don't stress. Unless your goal is to win mead competitions or please your friends, there's no reason to take the fun out of it. As a matter of fact, I would argue that if either of these are your goal, it's even more vital to please yourself first.

Table 4.2. Commercial Yeast Strains for Mead

Yeast Name(s)	Description and Recommendations for Use
Lalvin EC-1118/K1-V1116	Good all-around yeasts with similar qualities; ideal for dry meads, sparkling meads, late-harvest melomels, and restarting stuck fermentations; mead made with this should be given time to fully age.
Lalvin QA-23	Dry melomels; ferments to dryness at low temperatures.
Lalvin 71B-1122	Ideal for sweet and semi-sweet meads, melomels made with dark fruits (enhances aromatics), and cysers.
Lalvin ICV D-47	A white wine yeast that is good for dry and semi-sweet meads; ferments quickly, but often needs extra nutrients to feed on.
Red Star Premier Cuvee/ Pasteur Champagne	Both dry wine yeasts that are good for sparkling melomels and have high alcohol tolerance (up to 18 percent).
White Labs WLP720/ Wyeast 4184 Sweet Mead	Sweet mead and wine yeasts that are designed to leave 2 to 3 percent residual sugars and to draw out fruity flavors in melomels.
Wyeast 4632 Dry Mead	Good for . . . guess what? Dry meads!
Muntons Ale Dry Yeast/ Coopers Ale Dry Yeast	Good, affordable all-purpose yeasts. The beer snobs may not like them, but they're both great for any standard honey ale (with the addition of nutrients) or if you just want to make a small mead with brewer's yeast.
Danstar Belle Saison Dry Yeast/Danstar Windsor Ale Dry Yeast	Both work well for Belgian ales, fruity honey ales, bragots, or any beer you want to age for a while.
Safale US-05 Ale Dry Yeast/ Safale S-04 Ale Dry Yeast	English ale yeasts with low-to-medium alcohol tolerance (up to 8 percent). Good for honey ales.
Safbrew T-58/S-33	Good, robust ale yeast for bragots; able to tolerate alcohol levels of up to 11.5 percent ABV.

I know too many brewers and mead makers who have given up simply because they went overboard on the technical details and enjoyed neither the process nor the final product. Mead is Odin's gift. Give it time and Odin will speak through you.

You're a Viking—you can do this!

COMMERCIAL YEAST

While I wild-ferment most of my meads, there are advantages to using commercial yeasts from time to time. Fermentation should commence much quicker with commercial yeast, and you can do your initial ferment in a carboy rather than in a wide-mouth vessel (although I often open-ferment for a couple of days when using commercial yeasts). Also, certain commercial strains will help to initiate a strong, vigorous fermentation and are beneficial when striving for a specific flavor profile.

When making meads with a large amount of fruit, I rarely use commercial yeasts due to the high level of wild yeast already present. I have provided a list detailing different yeast strains and their uses in mead making in table 4.2. Once you get to know the different strains and the effect they have on your mead, you may become interested in combining yeast strains. I know mead makers who have done this to great effect, even following up a wild fermentation with a commercial yeast. You're welcome to experiment with this right off the bat, but I recommend getting to know your yeasts first by playing around with single strains and wild ferments individually, and learning about the flavor profile each imparts.

CHAPTER FIVE

Brewing the Drink of the Gods: Techniques for Making Wildcrafted Mead

When it comes to mead making, there are as many techniques and recipes as there are mead makers. I encourage you to develop a technique that works for you. You can learn from others, but it's more important to learn from yourself and trust your instincts. Before you know it you'll have developed your own "mead personality," and the meads you make will impart the essence that you put into creating them. Some meads, like delicate floral meads or light, effervescent fruit meads, will have a pleasant, alluring character that welcomes you to savor each sip while exploring the profile presented by the native yeasts and other ingredients. Some, like bragots, meads made with bitter herbs, or t'ej, will be aggressive and testy, and will seek to wrangle your taste buds into submission. Give each one a chance and don't fight it.

Start out simple and if it works, stick with simple. This doesn't mean you shouldn't experiment. Experimentation is what mead making is all about. Just be sure to approach each batch of mead with the intent of enjoying the sacred journey you're about to undertake, accompanied by all manner of tiny creatures who will speak to you and befriend you if you allow them. The mind-set in which you approach mead making can have a very real effect on the final outcome. Be mindful of your process and environment, but don't be overly regimented in your approach. If you get frustrated during any part of the process, stop, take a deep breath (or several), center

A 1-gallon batch of wild-fermented flower (wild violet, rose of Sharon, dandelion, and elderflower) mead happily bubbling away.

yourself, and get back to work . . . I mean, play. When trying out new techniques and recipes, I recommend small batches (1 to 3 gallons [4 to 12 L]). Less volume means less cost, less effort, and quicker gratification.

Preparing for—and creating—a wildcrafted brew is both mental and spiritual. Brewers I have spoken with employ techniques ranging from clinical to mystical. Perhaps this says more about the person than it does the process, but I prefer the latter approach. It allows me to communicate with my ancestors—to feel that I'm partaking in an ancient practice that connects me with the tiny creatures that surround us and are vital to creating a successful fermentation. For all I care, we could be talking about bacteria and yeast that can be seen under a microscope just as much as faeries and pixies frolicking playfully just out of the corner of our eye. As far as I'm concerned, they're one and the same. Before we had any way of knowing that yeast and bacteria even existed, we developed elaborate rituals to invoke the spirits that would bless our brews with boozy goodness. This may come as a surprise, but the beings I like to invoke when I'm brewing are Scandinavian in origin. Ancient European cultures (and many others) developed bonds with the wild yeasts they invoked that literally lasted generations. Once they had initiated fermentation, the Norse were careful to show respect to the beings that had infused their newly created brew with sacredness so that they would stick around. They did this by saving them on yeast logs and stir sticks, and in fermentation vessels—or by drying and saving the yeast-covered juniper branches they used to filter their brews.

In ancient Norse and modern Scandinavian languages, there are multiple words that share similar roots and reference the fermentation process in some manner. "The different regions of Norway named the thing that brings the ale into being *gjar*—'working,' *gjester*—'foaming,' *berm*—'boiling,' *kveik*—'a brood that renews a race,' *nore*—'to kindle a fire,' *fro*—'seed,' and one whose exact meaning is unknown—*gong*. . . . Once the gong or *bryggjemann* or kveik had come, the brewers and their culture had a special relationship with them. In many cultures, indigenous and otherwise, the wild yeast that came into the wort would be kept and nurtured as part of the family."[1] Being that the translation of *bryggjemann* is essentially "brewing man," I like to think of invoking this being when I'm creating a new brew. I then find ways of thanking the spirit that has blessed the concoction I have just brought into existence through various methods of passing along my new yeasty friends to future brews.

Harvesting Wild Yeast

The oldest domesticated living organism is not a horse or a chicken, nor is it corn or wheat. It is a wild single-celled, asexual creature capable of preserving food, making bread rise, and fermenting drinks. It is yeast.

—*Amy Stewart,* The Drunken Botanist[2]

Fermenting mead with wild yeasts offers myriad benefits to health and well-being. Honey itself is a natural preservative. In pure, concentrated form, it inhibits the growth of microorganisms, but is full of inactive bacteria and fungi.[3] If honey is diluted with water and allowed to wild-ferment—with care and attention to the process—the "good" microbes and fungi (that is, yeast) will thrive. Techniques such as boiling or pasteurizing the must will still produce a palatable mead, but the health benefits of the honey will be negated. By wild-fermenting at low to no heat, and adding herbs, spices, and other natural flavorings, you can create a beverage that is both nourishing and stimulating. This isn't to say that there aren't instances in which you might want to use commercial yeast or employ heat, but no-heat wild fermentation is just as viable an option as other techniques. To debunk a myth I've heard from folks who approach me about wild fermentation a bit warily, there are zero safety concerns when working with wild yeasts, and few issues regarding negative effects on flavor (provided the appropriate steps are followed). Mead has near-unlimited potential when you make it yourself, and there's no reason not to try out a wide variety of processes and ingredients. Wild fermentation is simply a method for creating alcoholic and non-alcoholic beverages, and one that has been around much longer than any other.

Over time, I've modified my technique for initiating wild fermentation and cultivating wild yeasts for future use based on experience and input from other wild yeast wranglers, but all in all it's a simple process with little room for error. The most crucial advice I can pass along about wild fermentation is that there is nothing crucial about wild fermentation. There are techniques you will develop over time as you learn about how to interact with the wild little creatures you're working with, but in the end you have to simply let go and learn from them. They will go where you want them to go if you pay attention to what they want from you.

I'll pass along a story that may help you understand this. When I was growing up on a farm in northern Kentucky, my family tried our hand at raising pretty much every kind of livestock. When I wasn't helping with the animals on our farm, I would do odd jobs for neighboring farmers. Sometimes I would be called in to help herd cows that had broken through a hole in the fence. When approaching such a situation, I would ask myself a simple question: "What do cows like?" Well that's easy . . . food! I would fill a bucket with sweet feed and go commune with the cows. The technique that I found worked best was to hold a bucket in front of the head cow, let her sniff at it and have a nibble, back up slowly, and proceed to lead the herd back to its home pasture. By anticipating the needs of these living beings and listening to what they wanted, I adjusted my behavior to accommodate them. They ended up where they were supposed to be in the end, and everyone was the happier for it. Since bacteria make up a very large portion of our bodies, we already commune with them on a daily basis. By paying attention to what the microbes in your environment want and tempting them with the proper nutrients, you can coax them to impart some of their magic into your mead in a similar manner as working with any other living creature.

My father, Wayne, made wine from the grapes and vegetables he grew, but when his winemaking equipment went unused for many years, I borrowed it for my mead making. When I told him of my wild-fermentation ways, he questioned his choice to let me use his equipment, asking if I was preparing for a long prison stay. Apparently he felt this was the only place where wild fermentation was worth pursuing. There is both a cultural and a generational stigma about brewing with wild yeast. But the more mead makers and winemakers I meet, the more I notice this trend shifting as people of all generations actively resist the mass sterilization that has become so entrenched in Western society since the early 20th century.

I am by no means anti-science; on the contrary, I find science and chemistry to be fascinating subjects. I do believe, though, that scientists and proponents of natural living can work together to learn from the mistakes of the past and harness scientific knowledge to better a society that has lost its roots. We should be utilizing our knowledge of bacteria and the building blocks of life to work *with* nature rather than continuing the failed policy of trying to conquer it. I truly hope to see the homebrewing community move away from the snobbery that prevails today, making it difficult for the average person to understand just how simple making fermented beverages can be. I am always amazed at the passion of the fermentation devotees I meet when I travel to skill-sharing and sustainable-living events. The flourishing and rapidly growing community of enthusiasts who have fully embraced the wonders of natural fermentation and are eager to share their skills is astounding and humbling.

PREPARING AN OPEN FERMENTATION

You may want to set aside some of the equipment listed in chapter 4 to use specifically for wild fermentation. Some would recommend doing this to avoid "contaminating" your other brewing equipment, but I've never been concerned about that. My reasoning is twofold: First, you might cultivate a colony of wild yeast you're particularly fond of and wish to pass down to future ferments; more important, you should develop a relationship with your wild-fermentation equipment.

You'll be using this equipment for mystical purposes, much like the ancestral cultures that called upon wild yeast and saved the yeast strains that befriended them. One technique for doing this was to use stir sticks that were passed down through generations as heirloom yeast strains. These sticks were referred to as magic sticks or totem sticks due to their mystical fermentation powers. Another method was to use a yeast log, often made of birch or juniper. The green log would be blessed by a shaman or other religious figure, runes or other symbols would be carved into it, and it would then be set out in a sacred place for the sap to dry out. As it dried, the log would develop cracks and crevices that, along with the runes, would provide a home for yeast when placed in the bottom of a fermentation vessel. The ancients, not knowing what yeast was, saw fermentation as a blessing from the gods. Gourds and other vessels were also held sacred due to the magic that was imbued in them with each fermentation. Clean your wild fermentation equipment lightly, but avoid sanitizer or boiling water. Keep it in a special place and commune with it regularly. For my equipment, I use a well-worn wooden stir stick passed on by my father from his winemaking days, along with one of his ceramic winemaking crocks.

Once you have your equipment and ingredients ready, the first thing you'll need to do is mix the appropriate amount of honey and water to create an environment ideal for calling in wild yeasts. Since you'll be adding more honey and water throughout the process, shoot for about a 60 percent to 40 percent ratio of water to honey, but don't worry about being too exact. For example, in a 1-gallon (4-L) batch for which you will eventually have about 4 pounds (2 kg) of honey to 1 gallon (4 L) of water, start with around 3 pounds (1.5 kg) of honey to a little under a gallon/liter of water. This process is based more on intuition and continual fine-tuning than on regimentation. That being said, always

PROTECT YOUR MEAD AT ALL COSTS!

Asgard, the home of the gods, had mighty walls built around it to keep out the giants, who were constantly testing the gods by trying to get past the walls. When making mead, particularly through open fermentation, the "giants" you'll need to concern yourself with are ants and fruit flies. Much of the winemaking literature states that a single fruit fly will turn a wine or mead into vinegar. While this may not be entirely accurate, you still want to keep them out of your mead. Ants can be pests but won't necessarily ruin a mead. I once left for a weekend and returned home to find that my cheesecloth covering had fallen into my mead and there were several dead ants floating in the mead. I shooed away all the living ants swarming around the crock, scooped out the cheesecloth and dead ants, and proceeded as usual. Each bottle of this mead I've opened has been excellent.

In my rush to leave the house, I had neglected to set traps. Rather than buying traps at the store, I keep ants away from my mead by setting out small containers with a bit of water mixed with honey or sugar and a tablespoon or two of cornmeal. The ants swarm around the container, ignoring my mead, and are gone within a day or two. What ants don't drown have likely returned to their homes with granules of cornmeal that proceed to expand in the stomachs of all who eat them. A bit of Borax (sodium borate) helps speed up the process. For fruit flies, I place some cider vinegar in some small containers, wrap plastic wrap tightly over the opening, and poke holes in the plastic with a fork. Fruit flies come in through the holes but tend to not come out. If you don't want to use plastic, place a couple of drops of dish detergent in the vinegar. This will create a film on the surface that the flies will stick to.

venture on the side of more water than honey for your initial ferment. This way you'll ensure that the must is liquefied enough to allow wild yeasts to enter and start partying with the enzymes and nutrients, and the yeasts will have less sugars to party with initially.

You may ask why you would want to deprive your yeasts of partying materials. Essentially, you have a stronger chance of a stuck fermentation if you give them too much sugar starting out. Think of it as dumping a bunch of starved Vikings fresh from a successful battle into a room with a seemingly unending supply of mead, roast goat, wild boar, and other victuals. Sure, they'll party hard most of the night, but eventually they'll have eaten, drunk, and fought themselves into a stupor. Start them out thinking they won't have much to feast on and they'll be more conservative in their imbibing. Once they're well satiated and have a good buzz going, you can bring in a couple more vats of mead and spits of meat. Then the party can commence and they can gorge themselves to their hearts' desire.

The same goes for yeast. Give them too much sugar and chances are they'll eat until they're bloated and then pass out (though if you have a strong yeast colony, things *could* turn out fine). Instead, start by starving them; once you've clearly got a strong fermentation going, add more honey (or if you like its current level of sweetness, a honey-water mixture) when you rack to a carboy. This way, you'll have the powerful combination of an active fermentation, additional aeration to get the yeast really worked up, and an oxygen-starved environment. At this point, they'll devour everything in sight until they're deprived of oxygen and pass out happily, leaving you with plenty of tasty booze. You can wake them up again throughout the process. Adding more honey, sugar, or other fermentables and aerating will get the party started again and result in an even more complex, boozy final product.

PREPARING A STARTER

Mead maker Patrick Ironwood lives with several generations of his family on the grounds of the Sequatchie Valley Institute (SVI)—a 300-acre land trust protected by a conservation easement through The Land Trust for Tennessee—where he brewed his first batch of beer at age 15 and has been brewing fervently ever since. Patrick's parents formed this 300-acre homestead located in a canyon bordering southeastern Tennessee's Sequatchie Valley in 1971, which—according to the Sequatchie Valley Institute's website—is now home to Moonshadow, an "education center and model of sustainable living."

When preparing for a batch of wild-fermented mead, Patrick likes to walk the 4-acre "edible landscape" surrounding Moonshadow, or the thousands of acres of wilderness in which the Sequatchie

The author demonstrates starting a new batch of wild-fermented mead. When possible, he likes to make mead outside near his garden and fruit trees to gather wild yeasts and botanicals for flavorings, tannins, and acids. Photo courtesy of Jenna Zimmerman.

Valley Institute is situated. Using his knowledge of plant lore, he'll pick botanicals that will instill his mead with a blend of acids, tannins, nutrients, and yeasts that are all interrelated due to the pollinators that have been busily visiting each plant. Among what he looks for are flowers that are just emerging and have therefore been recently pollinated, and fresh, whole fruit (nothing dried or overly blemished). What he's primarily seeking are the nutritional, medicinal, and magical benefits these plants will impart, which he refers to as the "trinity of

MARISSA PERCOCO
EMPLOYING RITUAL AND RHYTHM TO
INITIATE WILD FERMENTATION

Marissa Percoco has a process that reminds me very much of how many ancestral cultures made mead—through ritual, rhythm, and communion with nature. She has found that the cycle of the moon and time of year have a strong effect on the vitality of her fermentations. She swears that some of her most active ferments are the ones that she starts a couple of days before a full moon. "The yeast are in cahoots with the moon!" she told me. When she starts a new wild ferment, she takes a fairly small amount of mead and water (about a gallon), gets it warm, and takes it outside with her, "listening" for whatever plants are in the yard or woods where she is currently residing. She picks whatever edible botanicals she feels are needed for that particular batch, and brings it back inside, or leaves it out with cheesecloth over the vessel if the weather is right. Over the next couple of weeks, she'll add more honey, water, and botanicals as she moves it to larger and larger vessels until she has an active fermentation and is ready to rack to a carboy.

wholeness." Ingredients for flavoring won't come into play until he's ready to make a full batch of mead.

When he returns home, Patrick takes a quart or ½-gallon jar, adds water, and mixes in some honey, a bit of sugar, and some orange juice. He then covers the jar or puts an airlock on it and lets it sit. When he's ready to make a batch of mead about a week later, he'll have an actively bubbling, locally cultured "bug" he can use to kick-start a new batch of mead. He finds this technique particularly useful for starting large batches of wild ferments, and helpful in general for preventing stuck ferments. His tendency in brewing mead, beer, and wine is to make large batches to share with guests at gatherings on his homestead. Trying to get 30 or more gallons of mead to wild-ferment can be particularly challenging without a bit of help.

A jar of mead starter resting amid some of the wild botanicals in the author's yard that were used to help initiate wild fermentation.

 ## Ginger Bug

If you want to create your own wild-cultured starter, you can emulate Patrick's technique, or create a ginger bug using items easily obtainable from the local health food store. This bug is highly versatile and can also be used for making wild-fermented sodas, or to start other ferments.

INGREDIENTS

2 cups (500 mL) dechlorinated water

2 tablespoons (30 mL) organic cane sugar (any sugar will do)

2 tablespoons (30 mL) grated, diced, or chopped organic (non-irradiated) ginger with skin on

PROCESS

1. Mix everything well in a quart jar or other wide-mouth container, cover with a cheesecloth, and let sit in a warm, dark corner away from other open ferments.
2. Continue feeding with equal amounts of ginger and sugar daily and stir several times a day until you have a bubbling bug, which can take anywhere from two days to a week.

3. Use a cup of the bug as a starter for a 5-gallon [20-L] batch, or a couple of tablespoons for a 1-gallon [4-L] batch.
4. Keep the bug going for future use by adding more sugar, water, and ginger, or even a bit of actively fermenting mead.

Initiating Fermentation

Whether you use a starter (also known as bug, barm, or backslop), or ferment straight from the mead must, the most important thing you will need to do throughout the initial fermentation process is to aerate the must. In a wide-mouth vessel, the best way to do this is to take your magic stick and stir with wild abandon. Get a good whirlpool going and then reverse to start another whirlpool. Stir for at least five minutes, then cover with cheesecloth or other porous cloth—this allows wild yeast to populate the must while keeping flies and other critters out. Place the vessel in a warm (60–80° F [15–27° C]), dark room and stir several times a day. Since I work from home and my fermentation room is easily accessible, I stir at least four or five times a day, sometimes

To aerate mead must when open-fermenting, stir vigorously with a stir stick to create a vortex, reverse direction, and repeat for at least five minutes. This is particularly important for wild ferments.

moving the vessel around the house or taking it outside on warm days. I try to keep it out of direct sunlight, as sunlight can have a negative effect over time. When starting a bug, though, it doesn't hurt to put your starter jar in sunlight for a couple of hours. The warmth of the sun can do wonders for waking up slumbering yeast.

After about three to five days (sometimes sooner), the must will begin to bubble and fizz when you stir. Some meads, particularly those made with yeast-covered fresh fruit, will have a healthy amount of visible foam, making it abundantly clear you now officially have mead. Note that the air bubbles that rise when you first stir aren't necessarily a sign of fermentation. You'll know you have fermentation when you stop stirring, hold your ear to the vessel, and hear the yeast pixies chittering away at you. Modern science will tell you that this is the result of yeast turning the sugars in the honey into carbon dioxide and alcohol. I prefer to think of it as the slumbering yeast having woken up and thanking you for invoking the *bryggjemann* through your vigilant stirring and caring attention.

Yeasts begin reproducing exponentially during this process, cavorting with one another and engorging themselves on whatever sugar and nutrients they can find. It's a veritable yeast orgy. Most ancient cultures

When you have initiated a wild fermentation, you will see varying degrees of foam rise to the surface upon stirring, and will hear the ferment fizzing due to the release of carbon dioxide.

actually thought fermentation was a kind of boiling process, as the visual cues of fermentation are similar. As a matter of fact, *fervēre*—the Latin root of the word *fermentation*—means "to boil" or "to seethe with agitation or excitement."[4] It makes sense that a word that is also the root of *fervent* describes the fermentation process. An actively fermenting brew is an exciting thing to see. The frothing, foaming, fizzing concoction you just produced should get you pretty darn excited as well.

Regardless of how you feel about the technical details of what just happened, what is important is that you congratulate yourself. You're a Viking and you've just made mead! I told you that you could do this! This is where you thank the gods, do a little dance, or quaff a drinking horn full of fermented goodness. No matter how many times I make mead, I always feel a tingle of excitement when I realize I've brought yet another delicious mead monster to life.

Go ahead and loosely cover the vessel with a lid or thick towel to allow gases to escape but also to minimize contact with outside air, which is no longer needed for the process. You don't to need to rack right away, since the fermentation will be active for another week or two, which means there is no danger of the must turning to vinegar. The fermentation foam forms a protective layer that keeps out any acetobacter contaminants, primarily due to the ongoing discharge of carbon dioxide. Sometimes you may even see a thin white film, or pellicle, form on the surface. This is a good sign, and should be left alone. You have just introduced *Brettanomyces*, referred to by beer brewers as "Brett" and by vintners as "Flor" (flower). This strain of yeast likes to party on the surface, linking hands and forming a barrier to prevent oxygen from entering the must. An oxidative yeast that prevents acetobacter from turning your mead into vinegar, Brett is commonly found on the skins of fruit and (theoretically) spread by fruit flies.[5] It's safe to stir it in, but wait until you're ready to rack or drink the mead.

If your mead tastes good and you don't have the time, interest, or equipment to bottle it, you can stop here and just start drinking. It will only be mildly alcoholic, but will be full of nutritious probiotics from the wild yeast and other bacteria. Most ancient meads were drunk young due to the lack of airtight vessels for long-term storage. I like to make small batches in a 3-gallon vessel with a spigot so I can take samples to gauge how the flavor changes. It will likely be cloyingly sweet after initial

Some pellicle that developed on a mead that thus far has performed admirably in taste tests.

fermentation, so you may only want to drink a small amount, water it down, or make a mixed drink with some quality liquor if you don't plan on bottling. Some meads, "small" meads specifically, are designed to be drunk young, as they complete fermentation in a shorter period of time than other meads. I'll warrant that if you've taken the time and trouble to make mead, you'll want to age it. Proceed to the section on long-term aging on page 85 when you're ready. But before doing so, you may want to consider using some of the actively bubbling ferment to start your next batch of mead.

SAVING YEAST FOR A NEW STARTER

Once you've got an active initial fermentation going, you can begin taking advantage of the rhythm of continuous mead making. Ladle or siphon out about a cup's worth of young mead and add it to your next batch, stirring or shaking vigorously to fully incorporate it. You can also save some in a jar with a loose-fitting lid in the refrigerator or in a cool, dark corner, but only if you plan on using it within the next week or so; without regular attention, it will lose its usefulness as a starter. To keep it alive longer, pour it into a bottle or other narrow-neck

Tara Whitsitt

Saving and Nurturing Yeast Cultures

Tara Whitsitt of Fermentation on Wheels keeps a strain of "mother culture" alive in a jar shaped like a pregnant woman. "I brewed something wild two years ago," she told me when I interviewed her in 2014. "It was a pear cider, and it just went crazy. Something really insane got in and it was really, really strong . . . It was like a freak of nature yeast culture that appeared. So, I saved the sludge and the yeast by-product, and I added that to some champagne yeast that I've been saving from a packet for a year."

Fermentation on Wheels is a project envisioned by Tara with a goal of spreading awareness of food sustainability through fermentation workshops. To accomplish this, Tara petitioned for donations to convert a bus into a biodiesel-fueled traveling "creative kitchen and workshop space." The bus left its home port of Eugene, Oregon, in October 2013 for a tour of the United States. The project has a strong community ethos at its core and, as Tara put it, "runs on the fuel of the good people who love and support it." Tara works with many farms, homesteads, organizations, businesses, and people in seeing her project to fruition. Although initially intended as a one-year circumnavigation of the United States, the project extended from 2013 into 2015 because, as Tara noted, "turns out the USA is big."

container with an airlock and feed it a teaspoonful of sugar or honey every week or so. You can even keep a starter alive like this with a packet of commercial yeast.

Although you needn't go to any more effort than this in wildcrafted brewing, if you want to save a "pure" yeast strain, the best way is to first pour the lees (sediment from the fermentation process) into a wide-mouth jar with a lid, let the dead yeast cells sink to the bottom, and then carefully scoop off the top layer. You should see some yeast floating in the liquid in the middle of the jar. This is the "clean" living yeast that

Tara Whitsitt from the Fermentation on Wheels project provided this picture of the bottom portion of her mobile fermentation station. From left to right: peach habanero mead; sour ale with lactobacillus and yeast from a petri dish sample of Heady Topper Double IPA (provided by The Alchemist brewery in Waterbury, Vermont); sour ale with lactobacillus, Heady Topper yeast, and *Brettanomyces*; and blackberry mead. Photo courtesy of Tara Whitsitt.

you'll want to carefully extract; the yeast that floats or sinks is dead and could impart unpleasant flavors to future brews.

Another method for extracting and saving yeast is to take it from a bottle-conditioned ale, beer, or mead (either yours or that of a commercial brewery). If you choose one from a brewery, I recommend using an unpasteurized ale. I don't know of any meaderies that make unpasteurized mead. If you want to use a commercial, pasteurized mead, look for a bottle with a fairly recent date, as the yeast will still be alive and viable for fermenting. Most traditional

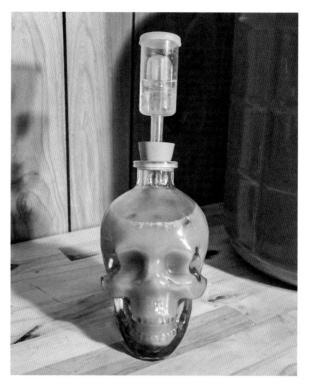

A convenient vessel for saving barm. Feed occasionally with sugar or active mead from a newly fermented batch.

Belgian ales are unpasteurized and bottle-conditioned, as are those of many microbreweries. Ale yeast is perfectly acceptable to use for some meads—specifically small meads and bragots. But try to avoid high-gravity beers, as the higher alcohol content can potentially cause off flavors due to its tendency to mutate yeast. In any case, open your brew of choice and carefully pour all but the last inch into a glass. What's left in the bottle is your yeast. If you want to be super-sterile about it, prepare a sanitized jar or glass, flame the tip of the beer bottle with a butane lighter, and pour the sediment into this prepared vessel. You can then use ale yeast to help out a wild ferment by simply pouring it into the fermentation vessel and stirring, or create a starter (also known as bug or barm) to save for later by pouring it into some warm mead must or beer wort, letting it sit for a couple of hours to aerate, pouring it into a bottle, and placing an airlock in the mouth of the bottle.

Saving yeast isn't just a homebrewing technique. Breweries and distilleries often reuse yeast, although breweries generally perform a yeast wash in a mild acid solution to eliminate potential contaminants, something you don't want to do unless you're really confident in your chemistry skills. In bourbon distilling, recently distilled (or "hot") sour mash is passed along to each subsequent batch. This "backslop" or "set back" allows for a continual cycle of proprietary yeast strains and is part of what allows each distillery to have a signature flavor. More important, by speeding up fermentation, this process minimizes the chances of foreign bacteria or yeast altering the intended flavor. This is how the old-time moonshiners did it and is a respected tradition, even in commercial distilling facilities. As with anything else involving brewing, you can get as technically sophisticated as you desire. Pick up a book on brewing chemistry if you want to pursue the fancy-pants route. I daresay there won't be any mention of brew pixies or the *bryggjemann*, though.

Long-Term Aging

Since continued contact with outside air can cause souring if the mead is left in an open-fermentation vessel, you'll want to rack into a secondary fermenter for long-term aging. In doing so, you'll be moving your mead from an aerobic ("with air") to an anaerobic ("without air") environment. Aging your mead in an anaerobic environment will result in a more refined flavor; additionally, you will achieve higher alcohol levels, as the yeast will eat up all remaining oxygen and sugars until there is nothing left for them and they return to dormancy in a yeast cake (or lees) at the bottom of the carboy.

The primary reason for racking to an anaerobic environment, though, is acetobacter, a bacterium that exists (in dormancy) in all alcoholic ferments, but particularly in wild ferments. Acetobacter requires continued contact with outside air to convert into acetic acid, which will turn your brew into vinegar. By eliminating the influence of outside air, you can allow your mead to focus on its job of becoming tasty booze. The best way to do this is to take a drilled stopper and place it firmly in the mouth of your carboy or jug and fit an airlock half filled with water

On Sourness and Funkiness in Mead

One of my least favorite early meads was one I had let sour a bit by leaving it exposed to air for too long (although some tasters said it was more tart than sour). As I was bottling it and labeling it COOKING MEAD, my wife tasted it and loved it. Several other friends did the same. After checking with their psychiatrists, or suggesting they look into one, I decided I could use this situation to my advantage. Now, if I happen to create a mead I don't care much for, there is always the opportunity to exploit my friends' tastes and offer some of my "very tasty" meads for trade. My point here is that if you end up with a mead that you don't like, you should generally be able to find a use for it.

Often it's not possible to completely eliminate a bit of souring, and a mead that tastes significantly of vinegar may as well be left out longer until it turns fully to vinegar and then bottled for cooking. Chances are, you'll want to drink your mead as well as cook with it. If the taste isn't too sour for your palate, you can always drink it young, perhaps blended with a bit of honey-water or other sweet drink. If you think it can still be saved, mix in as much honey water as you feel it needs, or blend with another mead that perhaps you find too bland. In ancient brewing (and in modern commercial brewing to a degree), blending soured meads, wines, and beers with younger, sweeter batches was a common way to "save" a batch, or was simply an expected part of the process.

Most important, don't give up just because you don't like the taste when bottling. Even if you enjoy mead with a bit of funkiness (that is, sour mead), bottling and aging it will enhance the flavor significantly. I've bottled meads with particularly strong levels of tartness or sourness and found that they mellowed out quite a bit over time. If you like mead with a bit of funkiness, drink it within a couple of months of bottling, as this is when stronger flavors will be more evident. Keep in mind that the higher the alcohol content, the longer the duration that sour, acidic flavors will last. You can also "back-sweeten" the mead with honey before bottling. See the troubleshooting section at the back of this book for tips on how to do this properly.

into the stopper's hole. Other options are to take a siphoning tube, fit it into the stopper's hole and put the other end into a bowl of water (you've just made a "blow-off tube," as referenced on page 138), or to place a condom or balloon over the carboy or jar opening.

RACKING AND FILTERING

Racking can be as simple as pouring the contents of a fermentation vessel into a carboy through a strainer or nylon mesh bag placed in a funnel. This works for transitioning to secondary fermentation, as it will strain out any large bits of flavoring material still in the mead (raisins, orange peel, et cetera) and provide aeration to invigorate further fermentation. However, if you choose to rack a second or third time to further clarify your mead, you'll want to employ a siphon to ensure the lees and any remaining bits of ingredients that have settled to the bottom stay at the bottom. This isn't absolutely necessary unless you're striving for something akin to the clarified wines and meads we've become accustomed to in modern times. To me, cloudy, yeasty meads taste just as good as if not better than clarified meads. When serving mead to company (the kind accustomed to fancy drinkin'), or when my wife and I want to sip on some mead with dinner, it looks much better in the glass when it's nice and clarified. However, when I'm drinking straight from the bottle or from my mead horn (unsophisticated as I am), I'm perfectly happy with mead that is full of cloudy, nutritious yeast.

There are a couple of different ways you can draw your mead out through siphoning. Homebrew stores usually have auto siphons, which allow you to siphon with a simple pumping motion through the use of a hollow racking cane attached to a siphoning hose and inserted into a tube to create a piston-type assembly. In my experience, these work great if everything is connected correctly, but if any air whatsoever is allowed to enter in where the hose connects to the cane, you'll lose suction. Even when clamping the hose to the cane, I've found these setups to be more frustrating than they're worth. I don't want to discourage you from using them, but my goal is always to strive for simplicity, so generally I just use my mouth to get the suction going. Now, don't get all high and mighty with me about sanitization. You're not going to ruin your mead by getting some germs on your siphoning hose. Using your mouth also gives you an opportunity to taste your

TRADITIONAL FILTERING PRACTICES

Filtering and straining of mead and beer, like most aspects of brewing in agrarian cultures, required a great deal of time, effort, and specialized equipment when made for sharing with the community. However, prior to the more complex techniques and equipment developed in Norway and elsewhere that were used in community brewing through the early 20th century, more rudimentary methods were used for thousands of years, and were likely what the Norse employed. As Odd Nordland notes in *Brewing and Beer Traditions in Norway*, "The practice of straining honey, milk, and ale or wine . . . belongs to the early stages of cultural development . . . In Finland in the past, the normal way of straining milk out on the pastures or on the dairy farm was to strain it through a funnel of birch-bark, which was filled with fresh fir or juniper twigs."[6]

In many ancient cultures, including early Scandinavia and Egypt, straw was twisted like rope and coiled to form a strainer. In order to test this method, I took some garden straw, hosed it down well, and set it out into the sun to dry. I then took a large plastic funnel, stuffed it with straw, and poured the contents of some mead must through it while

mead as it progresses, and to save a bit for putting into a hydrometer or vinometer (see the glossary or the *Testing Alcohol Strength* sidebar) to test its alcohol level.

Before racking by siphoning, start by positioning the vessel that currently holds your mead (or other boozy concoction) on a table or other platform above the vessel you'll be siphoning into. If you jostle the brew around a bit in the process, give it an hour or so to settle. Take your hose and carefully place it into the *from* vessel so that its end rests just above the sludge in the bottom, clamp it in place if you feel the need, stand up, and start sucking on the other end. Once you've got a steady stream, place the tube in the *to* vessel (you may also want to put a bit in a smaller vessel you plan on using for testing), and let it flow. If you need to stop the flow to move it between vessels, put your thumb over the opening. From here it should be a quick process.

racking. To be safe, I also put in a strainer that was designed to fit into the hole at the bottom of the funnel.

It turns out that the straw strainer was more effective, as the plastic strainer clogged quickly, causing me to lose some mead on the floor. When I pulled it out and just used the straw, I noticed that most of the bits of honeycomb and raisins stayed in the straw with only a small amount of residue making it into the carboy. Although I trust the source of my straw, it may not be the best route to take if you want to use vegetal matter as a strainer. Any type of edible, flavorful vegetation will do. Even today in some pockets of Scandinavia, brewers use juniper branches to filter homebrewed beer.[7]

Another ancient tradition that still exists in some cultures today involves drinking straight from the vat in which the mead or beer was fermented, with no transferring or filtering at all. This was (and is) a community ritual, with partakers sitting around the fermentation vessel in a circle. In addition to using ladles with strainers, participants simply suck through a straw made from the hollow stalk of a plant such as a thistle.[8] You can do the same using a siphoning tube if you want to drink a young mead without bottling or filtering, or just want to test the flavor before racking.

Watch the vessel you're transferring from and lower or raise the tube to ensure you get as much of the clarified mead as possible into the secondary container. But if you're okay with a bit of a yeasty flavor, don't stress about this too much. You can save the remaining lees for cooking, compost, or as backslop (along with a bit of must) for future mead. Racking is inevitably a messy process, so be sure to keep a moistened towel handy for cleaning up messes. Putting a funnel in the mouth of the carboy can also help keep mead from splashing on the floor.

Since racking will leave behind a bit of liquid, there will be space left in the carboy that you need to fill before you seal it with an airlock. When long-term aging, you should always fill the carboy up to the neck, only leaving an inch or so of space between the neck base and the carboy opening. If you feel your mead is sweet enough, mix an equal amount of honey and water and top it off. If you want it more or less

When racking, try to get the tube as close as possible to the bottom without touching the lees (sediment) on the bottom. Excess must and some of the lees can be saved as barm for future batches.

Melomels such as this elderberry melomel may need additional rackings to filter out the lees and other sediment created by the fruit.

Testing Alcohol Strength: Fancy Gadgets and Good Old-Fashioned Intuition

The various instruments available to us in modern times for checking alcohol levels and performing other tests look like they belong in a chemistry lab. Hydrometers and vinometers are cylindrical glass instruments that allow for precision testing of alcohol level (although a vinometer is less precise). The basic concept of the hydrometer has been around for a while. An unfermented mead, wine, or beer has a higher sugar content than it does later in the process. Because alcohol is less dense than water, a hydrometer can be used to measure the changes in gravity that occur through the various stages of the fermentation process as sugar is fermented into alcohol. By taking note of the difference from initial fermentation through bottling, you can determine the potential alcohol level of your final brew, usually designated as ABV (alcohol by volume). In early commercial brewing, this was done with large, complicated contraptions. Prior to that, the progress of a fermentation was monitored through more intuitive means.

For most of my brewing, I use intuition. I have hydrometers and have used them for making beer but have since weaned myself off brewing "to style" or shooting for a specific alcohol strength. Rather, I prefer to watch and "listen" for my mead, wine, or beer to let me know when it's ready. Intuition needn't come from years of experience. You can develop it based on your own experience from the get-go. I do, though, find a vinometer handy to have around. It's not entirely accurate for anything but a dry wine or mead (sugars and other ingredients can throw off the measurement), but it's useful for obtaining an approximate measurement. For clear meads, add a drop of food coloring to make the measurement easier to read. Homebrew stores also sell refractometers for measuring alcohol content digitally if you're a gadget geek.

sweet, adjust accordingly. The goal is to minimize the amount of air in the vessel. Even with an airlock, too much air in the vessel can result in acetobacter souring the mead.

A second racking of a chai metheglin to siphon the mead off the lees (sediment) it has created through the fermentation process. This helps enhance the clarity and flavor of the final product.

Once you have racked your mead, patience is the key. This may be easy if you're coming at mead making from a winemaking background, as wine can also require lots of time and several rackings. I initially brewed beer and was accustomed to a quicker turnaround time between racking and bottling. While there are types of mead that are okay to bottle sooner than others, a good rule of thumb is to wait *at least* six months before bottling, although one year is best. Honey has a lot of sugars in it, which can take some time to break down. Until it is fully fermented, mead will continue to have residual carbonation.

Racking into a plastic bucket with a spigot in preparation for bottling. It's best to not use plastic buckets for long-term fermentation.

BOTTLING

Other than time, the best gauge for determining when a mead is ready to bottle is to simply watch it closely for a few minutes. Watch for carbonation bubbles, and monitor the airlock for any sign of residual carbon dioxide in the carboy. If there is minimal to no sign of carbon dioxide in the airlock (occasional bubbling or a deflated balloon or condom, depending on your method of keeping air out), it's likely almost ready for bottling. It's still a good idea to give it another couple of weeks to ensure that fermentation has fully stopped.

When first trying out mead making after some experience with beer, I was too eager to bottle. In one instance, I was woken up at four in the morning by corks popping in my kitchen. Other times I walked into my cellar only to find that it smelled suspiciously delicious, subsequently noticing a cork or two—and in some instances bits of glass—on the floor. Cyser (or any mead with fruit) requires even more patience because you're adding additional sugars to the mix. I once let a cyser sit in the carboy (after a couple of rackings) for nearly a year. It seemed to have finished fermenting, so I bottled it. After a couple of weeks, I opened a bottle to test it and was met with a geyser of cyser. Not wanting to let this hard-earned batch of cyser turn my cellar into a war zone, I took the bottles and sat down in the living room, planning on opening them slowly and pouring them into a fermentation bucket. After making a mess of the living room—and getting some looks from my wife—I decided to move to the back porch, considering that it was a nice summer evening anyway. After a couple more months, it was finally ready for bottling. My cyser was saved, and who knows what else was.

The author demonstrates bottling mead from a bucket with a spigot.

Frustrating as instances like these may be (and believe me, they are frustrating), it's important not to give up. Granted, the hectic schedule of modern life can make it difficult to devote tons of time to brewing, but after you've gone through a bit of a trial and error, brewing needn't be a time-consuming, aggravating process . . . unless of course you're like me and can't help experimenting further just when you've got things figured out. As with any other aspect of homebrewing, I recommend cleanliness when bottling, and sanitization only if you're comfortable using chemicals. When I prepare to bottle with corks, I get a pan of water boiling, cut off the heat, give it a couple of minutes, and then place as many corks as I think I'll need (plus a few) in the water. This will sterilize the corks and will help soften them.

BOTTLES

Don't feel like you should go out of your way to spend money on bottles. You can save bottles yourself, ask friends to save them for you, or raid the garbage and recycling bins of restaurants or bars. The best bottles to use vary depending on the type of mead you are making, how long you plan on aging it, and how much you plan on drinking in one session.

Jugs

The simplest method for bottling by far is to ferment your mead in ½-gallon (2-L) or gallon (4-L) jugs, cap or cork the jug when fermentation is complete and there is minimal to no carbonation (after racking it off the lees), and drink it when you're ready. However, once you've uncorked the jug, be sure to finish it within a week or the extra air in the jug will oxidize and sour the mead. If this happens, do what the old-timers did and add a bit of honey or new mead to it before drinking.

Screw-Top Bottles

One type of bottle you won't often find recommended for homebrewing is the screw-top. To be fair, there are valid reasons why brewers generally avoid screw-top bottles, primarily:

- They're made with thinner glass and therefore break easily when capping, particularly with a wing-style capper.

- If you bottle carbonated mead or beer, the thin glass can lead to an increased likelihood of bottle bombs.
- The threads on screw-tops aren't as reliable as the lip of a pry-off bottle for keeping outside air in and inside air out.
- They're not pretty and imply cheap contents.

I call hogwash on the last point. While cheaper beers and wines do often come in screw-top bottles, lower price doesn't always mean lower quality. Also, for the frugal homebrewer, it's not what's outside, but what's inside that counts. The other points are valid, but don't let them stop you. I admit I've always used cork-ready and pry-top bottles. However, I've heard of many a homebrewer who has bottled with screw-tops to great success, and have had my share of thick-glass bottles break at the neck when capping or uncapping.

Thick-Glass Bottles

For meads with any degree of carbonation, always use swing-top (also known as flip-top or Grolsch-style), champagne, or Belgian-style beer bottles. All of these have thick glass and are designed for holding highly carbonated beverages. An additional advantage of champagne and Belgian-style beer bottles is that they have a lip under the opening that can be used to wire down a cork to prevent it from popping prematurely. Flip-tops are by far my favorite bottles to use. They're thick and sturdy, the cap is always there, and the rubber gasket that acts as a seal will last a long time and is replaceable. You can use these for pretty much any style of mead or beer.

Corked Liquor Bottles

Several of the finer bourbons, whiskeys, and Scotches are corked with a "T-cap" cork that can easily be pushed in and pulled back out. Somehow I manage to keep a fair share of these bottles around. They're great for bottling mead because they're simple to use, pre-sanitized (they've been soaking in high-proof alcohol for who knows how long), have thick glass, and the cork can be used to "stopper" a bottle for finishing later. If you use them for a sparkling mead, be sure to secure the cap in some fashion to keep it from ejecting prematurely and ruining all the fun. If nothing else, a rubber band or non-galvanized (in other words, non-poisoned) wire can be used to hold it down.

A Note of Caution About Bottling

Even with thick-glass bottles, please take precautions when storing and opening any alcoholic beverage. Unless you're shooting for a sparkling beverage, be sure you have given whatever you're brewing enough time to decarbonate before bottling. I've bottled beer and sparkling mead at the appropriate time—in thick-glass bottles—and still found that some of them exploded (in most cases, it was simply a case of the bottom of the bottle having given in to the pressure rather than a full-on explosion).

When opening an untested homebrew, I tend to do so outside—carefully, and with a towel wrapped around the bottle. This way I'm prepared for any messes. For the most part, I haven't had any major problems, but I've heard horror stories of bottles exploding and leaving shards of glass in the wall, and causing messes in rooms that have led to domestic disputes. This isn't to say I haven't had my own experiences . . .

After I had a few homebrewed beers under my belt, I happened to open a bottle in a small cabin with a very low kitchen ceiling. The resultant geyser decorated the ceiling, the sink, and a good bit of the kitchen with malted boozy goodness. Fortunately I was single and living alone, so no relationship issues ensued. After I'd had more than one bottle from this batch open this way, I took a couple to a party, insisted that everyone come outside to watch me open a bottle of beer, and then listened to the crickets chirp as the bottle opened with nary a sound. The lesson here is that even within the same batch, carbonation can be unpredictable.

CORKING AND CAPPING EQUIPMENT

While there are various DIY methods for making bottle cappers and corkers, there isn't much reason to. They're fairly easy to find and come in a range of prices (and quality levels). Most are lever-based and therefore fairly ancient technology, so you can still feel like you're brewing like a Viking.

For both corking and capping, the cheapest devices you'll find are handheld. I used a handheld, wing-style bottle capper for years with no problem, but didn't like the wing-style wine corker I tried one bit. A winemaking friend gave it to me and asked that I not give it back,

A sampling of bottles, corkers, and cappers. From left to right: a small lever-style capper, a plunger corker, a wing-style corker, a versatile floor corker/capper, a Belgian- or champagne-style bottle with a wired-down cap, and a wing-style capper. Behind the wing-style capper are a capped bottle and a swing-top/Grolsch bottle.

which made sense once I used it a few times. I nearly knocked a couple of bottles over and no matter how I adjusted it, I could never quite get the cork all the way in. I highly recommend a floor corker. Most of the quality ones you can find are made in Italy. A standard Italian floor corker can run you up to $150. While I appreciate the mostly metal construction and vintage look of these corkers, my wife purchased me a combination corker/capper that is pretty much all I use anymore. For around $75, this adjustable machine allows you to cork and cap bottles of almost any size or shape. It's made out of mostly plastic, which makes it very lightweight for its size. Other styles of corkers available include

The author demonstrates capping a bottle with a wing-style capper. Behind him are caps, bottles, a bottling bucket, and carboys of mead in various stages of fermentation. Photo courtesy of Jenna Zimmerman.

compression corkers, handheld plunger corkers, mini corkers (which should be used along with a rubber mallet), and plastic-plunger corkers. Take note, though, that most of these alternatives don't handle #9 corks well, which I will discuss in the following section.

Caps and Corks

If you bottle in a beer bottle with a capper, you can use standard crown caps or oxygen-absorbing (also known as oxygen-barrier) caps. The insides of these caps are lined with an oxygen-permeable film layer that absorbs any oxygen present in the bottle's head space. When aging mead in wine bottles, there are a few things you'll want to take into consideration when corking. If you plan on aging some of your mead for several years (which you should), it is critical that you choose corks of the proper length and diameter. You want to allow in minuscule amounts of air to help your mead age slowly, but not so much that it causes flavor deterioration. Boulder, Colorado's Redstone Meadery states on their website that they do not cork, in order to avoid oxidation. Instead, they bottle their mead in swing-top bottles, as they feel it shouldn't breathe while cellaring. They also don't boil their must and use no sulfites due to their self-described natural "philosophy of mead." Redstone is one of my favorite meaderies, so I see no need to decry these tactics. Another advantage swing-tops provide is that you can (very) occasionally open a bottle for a test swig, and close it again if you feel it isn't quite ready. It takes time and money to build up an inventory of swing-top bottles, though, and corking is a time-tested technique for bottling mead, so don't feel like you should invest in large quantities of swing-top bottles.

When you purchase new corks, the label on the package will have some numbers on it. The most common are: #7X 1¾ inches (4.4 cm), #8X 1¾ inches (4.4 cm) and #9X 1¾ inches (4.4 cm). The first number (between the hash mark and the X) is simply a size designator (like S, M, and L), and the second is the length, which in these three examples is the same. The vast majority of wine-bottle manufacturers produce bottles with an industry-standard opening (¾ inch [2 cm]), so depending on the size of the cork, more or less air (the larger the first number, the wider the cork) will get into your brew over time.

My suggestion is to always use #9 corks, unless you set aside a few bottles that you definitely plan on drinking within a year or two. Be sure to label these bottles appropriately. It's generally best to avoid #7 corks or tapered corks (which can also pop out due to carbonation or incomplete fermentation) for anything you don't plan on drinking within a couple of months. I always keep plenty of #9s on hand, even if I plan on drinking a bottle or three early. As a rule of thumb, use #9 corks for aging mead two to seven years, #8 corks for aging one to two years, and #7 corks for aging three to six months.

If you want to set aside a reserve for 10 years or more, you can go online or ask at a homebrew store for corks made from superior-quality materials designed for that purpose (or try your luck with a swing-top bottle). In talking to other mead makers, I've heard varying opinions on just how long a homemade mead can be saved. Some have told me that they feel their mead decreases in quality after 5 years, and others have said that their mead continues to improve after being aged for more than 10. I once talked to a beekeeper who said he opened a 50-year-old bottle that tasted phenomenal. Personally, I tend to keep most of mine for two to five years, occasionally sampling some early and taking notes to compare how a batch is aging. I do, though, keep a reserve stash that I plan on saving for as long as my patience will allow.

SPARKLING MEAD: A PRIMER ON PRIMING

For most of my meads, I like a bit of sparkle. Perhaps it's because I'm a beer lover at heart and have grown accustomed to carbonation in my alcoholic beverages, or maybe it's simply that I enjoy my mead with a bit of liveliness. This is where your personal preferences will come into play. If you prefer your mead to be flat, then by all means bottle it flat.

If you want it to be carbonated, you can bottle when there is still a bit of residual sugar. To do this, bottle when you still see a small amount of bubbling in the carboy, or when your airlock burbles once every minute or two. You can also take a bit of mead out with a turkey baster and swish it around in a glass to see if it has a bit of fizz to it. Finally, you can take a tablespoon or two of sugar, sprinkle it in the carboy, and watch for a reaction. If the reaction is strong, this means the yeast is still working and you will have carbonated (possibly overcarbonated) mead. Giving it more time or adding 1 cup of sugar or honey (no more than 4 cups) over a couple of weeks will allow the yeast to finish their feeding and thus eliminate any residual carbonation.[9] These methods aren't always reliable, unless you've developed an instinct for knowing when to bottle. A safer and more consistent method is to wait until the mead is fully fermented and then prime either some or all of it prior to bottling.

Priming is simply adding sugar for carbonation just before bottling. Any type of sugar will do, but honey is best. Wait until bottling day to do this. For a 5-gallon (20-L) batch, take 2 cups of warmed water and dissolve ½ cup (120 mL) of honey or ¾ cup (180 mL) of sugar in it. Stir well, add it to your mead, and stir the mead slowly to mix the sugar in without overly aerating the mead. If you do this with 1 gallon (4 L) of mead, or would like only some of the bottles to be sparkling, add around 2 tablespoons (30 mL) of honey or sugar per gallon/liter. The sugars will interact with the remaining yeast in the mead to produce a mead with a nice, bubbly sparkle. If you have allowed the mead to fully ferment before doing this, you're safe using a standard wine bottle. To be extra safe, or if you want to add more sugar for a champagne-like effervescence, use champagne, Belgian ale, or flip-top bottles. Although you should wait longer before drinking most meads, the bottles will be fully carbonated in about two weeks.

CHAPTER SIX

Basic Mead Recipes and Some Variations

Top-quality mead ingredients and a poor recipe beat poor ingredients and a top-quality recipe nine times out of ten.

—*Ken Schramm*, The Compleat Meadmaker[1]

For each of these recipes, it's important to understand the overall mead-making process as outlined in chapter 5, as you will inevitably need to make adjustments. I've noted any deviations from the standard process, but you'll need to use common sense as you develop your own technique. For instance, I don't always list the full amount of water, but rather the recommended amount for creating the must. When you add the must to the secondary fermenter, you will likely need to add more water or honey-water to reach the neck of your carboy. Sometimes it is best to wait until after fermentation slows down—for instance, when you're making a fruit mead with an aggressive fermentation. Also, although I will sometimes recommend botanicals for tannins and acids, be sure to read through the discussion on when, how, and if to add these in chapter 4, and have a look at table 4.1, Botanicals Containing Acids, Tannins, and Nutrients.

I encourage you to adjust ratios and ingredients to make mead you can call your own. A word of caution: You can get some great info and recipes from homebrewing websites and forums. The Internet, though, provides a certain degree of anonymity and is full of opinions that may or may not be based on actual experience or well-researched knowledge. Take *everything* you read with a grain of salt, including my own articles. When viewing my articles, scroll through the comments for any updates in my process or lessons learned. You can learn a lot from viewing and participating in forums on mead making, but you can also waste a good bit of time. Always double-check anything you learn, and take careful notes when trying something new.

You can substitute fruits, spices, and other ingredients with what you have available or leave out a particular ingredient if you don't think you'll like the flavor it will impart. If the ratio of honey to water would result in say, a semi-sweet mead—and you would prefer a drier or sweeter mead, adjust accordingly. Above all, seek out quality ingredients, and don't feel obligated to follow any of these recipes exactly. Many of them I have made with success; others I have obtained from trusted mead-making associates, books, or Internet sources. Work with what you have available and can source locally, sustainably, and naturally. Most of all, make small batches when trying a new recipe, taste throughout the process, and have fun!

When making mead, even when going strictly from a recipe, it's always good to take detailed notes. Since the overall process can take up a fair bit of time, you may forget what you did earlier on. Write down specific ingredients, ratios, measurements, thoughts, variations, and anything else that comes to mind. Continue by marking down the date of when you initiated fermentation, each time you racked, and when you bottled. Noting how the mead tastes each time you open a bottle doesn't hurt, either. With that being said, I did a pretty lousy job of this until I began writing about mead making. I usually started with a recipe and adjusted using my intuition. More often than not, I turned out perfectly fine brews. But I take notes religiously now, and highly encourage you to do so. These can be useful for repeating batches you like, troubleshooting those you don't, and having a record to pass along if you enter your mead in competitions—or if a discerning friend tastes some and wants to know what is in it.

Show Mead

For the purpose of mead competitions, the term *show mead* was developed to denote a mead containing only honey, water, and yeast (although acid and tannin adjuncts are generally allowed provided they aren't used in excess). Some purists will try to tell you that a traditional mead must only contain honey, water, and yeast, and will even go as far as to not consider anything else to be a mead. If their definition of *traditional* goes back to the first meads that were spontaneously fermented by accident, they may have a case. But even so, when Neolithic man first discovered mead, it could very well have been from letting water get into honey in a container that also contained residues of berries and herbs they had gathered. Archaeological, written, and anecdotal evidence proves that meads have been flavored with a wide range of ingredients from Neolithic through modern times. What I'm getting at here is that the "traditional" argument has no merit. Mead can—and should—be made from all manner of ingredients, although there is something to be said for the delights of a well-balanced pure honey-water mead.

I recommend making several small batches of show meads using various types of honey and ratios of honey to water. This will help to fine-tune your personal technique, and will prepare you should you decide to enter a competition. Although competitions have their place and can be a lot of fun, I personally don't look at homebrewing as a competition. Just as with beer competitions, mead competitions tend to focus far too much on style. Style is a very subjective thing and had little place in ancient brewing. I like to use styles as a springboard for customizing my own brews, and tend to avoid thinking of whether or not I'm brewing "to style," as I feel that limits my ability to enjoy myself and create something truly unique.

Semi-Sweet/Medium Show Mead

This is my workhorse mead. Nine times out of ten, people who don't like dry or sweet meads will like a semi-sweet. It has something for everybody, and is a good mead to prepare for sharing with large groups, particularly folks who haven't yet been initiated into the world of mead.

INGREDIENTS FOR 1 GALLON (4 LITERS)
3–4 pounds (1.5–2 kg) honey
Enough water to fill a 1-gallon (4-L) jug to its neck
Wild yeast, barm, or sweet wine yeast

10–12 raisins, one black-tea bag, bark tea, or other tannin/
    nutrient of your choice
1 teaspoon (5 mL) yeast nutrient if you don't trust the raisins to
    do the job
Acid to taste

Sack/Dessert Show Mead

A mead for the sweetheart in all of us. Be sure to choose a quality honey with a flavor you enjoy, as the honey will come through strong. It will inevitably be flat; don't even try to carbonate it. Savor it slowly after dinner or with a platter of fruit, crackers, and cheese. It is also excellent fortified with a strong liquor.

INGREDIENTS FOR 1 GALLON (4 LITERS)
4–6 pounds (2–3 kg) honey
Enough water to fill a 1-gallon (4-L) carboy to its neck
20–25 raisins, one black-tea bag, bark tea, or other tannin/
    nutrient of your choice
2 teaspoons (10 mL) yeast nutrient if you don't trust the raisins
    to do the job
Wild yeast, barm, or sweet wine yeast
Acid to taste

Dry Show Mead

The less-is-more mantra holds true here. Although you won't get the overwhelming mouthfeel of a sweet mead, the subtle flavoring of the honey in a dry mead is worth searching for with every sip. Be sure to taste throughout and adjust acids and tannins as needed, as it can potentially be a bit insipid. I bottle most of my dry meads a bit early or prime the bottles for extra sparkle.

INGREDIENTS FOR 1 GALLON (4 LITERS)
2–3 pounds (1–1.5 kg) honey

Enough water to fill a 1-gallon (4-L) carboy to its neck
Wild yeast, barm, or dry wine or champagne yeast
10–12 raisins, one black-tea bag, bark tea, or other tannin/
    nutrient of your choice
1 teaspoon (5 mL) yeast nutrient if you don't trust the raisins to
    do the job
Acid to taste

SMALL MEAD
(OR MEAD FOR LAZY AND IMPATIENT VIKINGS)

Small meads (also known as short meads or quick meads) are ideal for beginning mead makers, or for trying out flavoring experiments in planning a recipe for long-term aging. They're also quick and easy to make using leftover honey and ingredients from a larger batch. They require less honey than most standard meads and can be either wild-fermented or fermented with ale or bread yeast. Due to the small amount of honey and the size of the batches, small meads are ready to drink much sooner than most meads. You can let them age for longer periods, but the point is to drink them young, sweet, and bubbly. I'll outline the basic ingredients and process, along with some suggestions for flavoring. Feel free to experiment.

 # Basic Small Mead

INGREDIENTS FOR 1 GALLON (4 LITERS)
1–2 pounds (.5–1 kg) honey, any variety
Enough water to fill a 1-gallon (4-L) carboy to its neck
Wild yeast, barm, bread yeast, or ale yeast (wine yeasts are
    designed for longer fermentation)
10–12 raisins, one black-tea bag, bark tea, or other tannin/
    nutrient of your choice
1 orange or lemon, ½ cup (125 mL) orange juice, or
    ¼ cup (60 mL) lemon juice (add to taste after
    fermentation commences)
Suggested flavoring additions: 1 cinnamon stick, 2–4 whole
    cloves, 1 whole nutmeg, 2–4 thin slices ginger, etc.

PROCESS

1. Mix the honey with the room-temperature water in a wide-mouth vessel and wild-ferment; or add one packet (2 teaspoons [10 mL]) yeast, barm, or other starter.
2. If you're wild-fermenting, wait until the ferment is active and pour it into a 1-gallon (4-L) jug with a funnel, or pour in fresh must and add yeast or barm.
3. Aerate the must by placing a lid (tightly) on the jug and shaking it vigorously, or using a thin stirring implement such as a chopstick to stir it vigorously.
4. Add the tannin and flavoring ingredients.
5. Airlock via your preferred method.
6. Taste a week or so after fermentation to determine if acid is needed and add to taste.

You have a few options for aging. You can wait about a month after fermentation commences to drink it. If doing so, handle carefully to keep the sediment (lees) that has settled to the bottom from mixing in (although lees has plenty of nutrients to impart), and pour or siphon into a drinking vessel. It will be mildly alcoholic, a bit bubbly, and plenty flavorful. Or you can siphon or pour it into some champagne bottles (which you'll need to cork), flip-top bottles, or even 2-liter plastic soda jugs, and put it in the refrigerator for 3 to 10 days. Handle sparingly and carefully, as pressure will be building. The refrigerator will slow down fermentation, though. You will have a sparkling, effervescent beverage if you go with this method. Drink it straight, or fortify it with your favorite liquor to make a cocktail. I recommend bourbon, Scotch, or brandy, but feel free to experiment to your heart's desire.

Skald's Verse-Bringer Spiced Orange Mead

Joe Mattioli's "Joe's Ancient Orange Mead" was an inspiration for this recipe. The recipe was passed along to me by some friends who had made it with great success, and is very popular in mead circles. Joe's goal was to emulate how the ancients would have made mead. As evidenced by this book, though, there is much more to making mead by truly ancient methods. Essentially, it's a small mead prepared with bread yeast. I've modified the recipe and technique a bit, but at its core it remains Joe. Note that you can also wild ferment this in an open-mouth vessel and then move it into the jug the recipe calls for.

INGREDIENTS FOR 1 GALLON (4 LITERS)

2–3 pounds (1–1.5 kg) light, floral honey
1 gallon (4 L) water
1 organic orange
1 teaspoon (5 mL) or 1 packet bread yeast or 2 tablespoons
  (30 mL) barm
Spices (2 cinnamon sticks, 2–3 cloves, 1 crushed nutmeg seed,
  3 crushed allspice seeds)
Ginger (2–3 ounces [57 to 85 g] fresh grated or 3–4 thin slices;
  optional but recommended)

PROCESS

1. Empty the water into a pot and heat to around
   70° F (21° C).
2. Mix the honey with the water until it is well dissolved and
   pour the must back into the water jug with a funnel to about
   2 inches (5 cm) below the neck.
3. Slice the orange into eighths and remove the rind from all
   but two or three slices to avoid an overly pithy flavor.
4. Add yeast or barm and all the spices, put the lid on tightly,
   and shake the jug like you mean it.
5. Place an airlock and cork in the opening, or cover with a
   balloon, condom, or plastic wrap held on loosely with a
   rubber band.
6. Store as you would any other mead.

I recommend racking at least once, but this isn't a mead designed for bottling (although it certainly wouldn't hurt to bottle it and age for a couple of months). After about a month, start tasting it. When you like it, drink it.

T'ej (Ethiopian Honey Wine)

When I first became interested in how the Vikings might have made mead, the simple, wild-fermented traditional process for making t'ej[*] (the national drink of Ethiopia) struck a chord.

T'ej is traditionally wild-fermented, but some commercial breweries (and homebrewers) will add yeast when they make it. Although you can make it without *gesho*—its core ingredient—I highly recommend locating some.

Gesho (*Rhamnus prinoides*) is a species of buckthorn native to eastern and southern Africa. If you live in or near a metropolitan area, try looking up an Ethiopian restaurant or grocery to obtain some gesho. Ask for some Ethiopian honey while you're at it. The honey traditionally used for t'ej is thick and creamy due to the comb that is mixed in with it. Since I live in a college town, I simply asked one of the Ethiopian students I knew if she could procure some when I heard she was going home for a visit. Lo and behold, she brought me back a grocery bag full of gesho twigs and leaves. T'ej is commonly made in the home (my benefactor said her mom fermented t'ej in the kitchen on a regular basis), and is a tradition Ethiopians like to carry with them when they leave the country.

 ## Traditional T'ej

Ethiopian food can be very spicy, and this sweet and tangy beverage is often served with meals to counteract the spiciness. Be sure to give t'ej a try if you're into spicy food. What follows is what I have found to be the most common way of making it traditionally. Details such as the ratio of honey to water, the way it is fermented—and other minor factors—vary, but as with any mead, you can simply experiment until you find a technique and a recipe that work for you. And if you can't get any gesho, don't let it stop you from making t'ej. You can make it with no bittering agent at all; with oak bark, twigs, and leaves; or with standard beer hops.

---

[*] You may also see the word spelled *tej*, *t'ejj*, or *tejj*. The former is simply a shortened, Anglicized way of spelling it. The apostrophe is meant to convey a sort of spitting sound between the *t* and the *e*.[2] I've generally heard it pronounced in English as *TAI-j*, *TE-zh*, or *TE-dj*.

INGREDIENTS AND EQUIPMENT
FOR 1 GALLON (4 LITERS)

1 gallon (4 L) water

3 pounds (1.5 kg) honey

4–5 gesho sticks (1–2 inches [2.5–5 cm] each), or
   alternative bittering/tannin ingredients

10–12 raisins

1 open-mouth vessel (1–3 gallons [4–12 L])

1 stir-sick long enough to scrape the bottom of your vessel

Cheesecloth, muslin, or a dish towel

1 funnel with a fine-mesh strainer, or strainer with
   a handle

PROCESS

1. Mix the honey with the water at room temperature or
   slightly warmed, continually stirring and scraping the
   bottom of the vessel until the honey is fully incorporated
   into the water.

2. Add the gesho or alternative bittering/tannin ingredients
   (such as oak bark or black tea). Or you can make 2 cups
   (500 mL) of tea from these ingredients and add after the
   must has cooled to 60–70° F (15–21° C).

3. Cover the vessel with cheesecloth, muslin, or a towel and
   place it in a warm, dark corner.

4. Stir several times daily to aerate and incorporate yeast from
   the air. Some recipes tell you to place a lid on the vessel and
   stir once a week, letting the natural yeast from the gesho
   initiate fermentation, but I like to aid the process with
   aeration and wild yeast from the air.

5. Whichever method you choose, you will likely see fermentation commence in three to five days; you'll know when
   you see and hear fizzing after stirring. If you notice mold
   growing on the surface of any of the ingredients, carefully
   remove it with a spoon, or simply stir it back in, as this is
   just a part of the fermentation process.

6. After about two weeks, remove the sticks and bark and
   strain out any residue.

7. Let it sit for another two to three weeks in the open-mouth vessel, stirring regularly, or transfer to a narrow-neck jug with an airlock.
8. Drink it now, or follow my aging suggestions below.

The reason I recommend straining it and transferring it to a narrow-neck vessel with an airlock once it has started fermenting is that my t'ej often develops a strong sour flavor when I've aerated it for too long. Some people like this and consider it a characteristic trait of t'ej. My mother-in-law, Babs Wright, tried some of my soured t'ej and said it tasted similar to the t'ej she'd had in Ethiopia; others who have tried the t'ej I made this way liked it. If you would rather have minimal to no souring, go ahead and transfer it to a narrow-neck jug with an airlock or any vessel with a loose-fitting lid (be sure to "burp" it every couple of days to release built-up pressure).

A convenient vessel for fermenting and sampling
small batches of mead, and for brewing t'ej.

I recommend tasting bits of it throughout this process to gauge how the flavor is progressing. I like to make mine in a glass vessel with a spigot so I can sample as I please. I tend to find these in grocery stores for fairly low prices.

If you like the t'ej while it's still sweet, tangy, and low in alcohol, drink it now. If you want to give it time to age, mellow, and dry out to something more akin to a dry mead, proceed as you would with any mead. Just be sure to add more water or a honey-water mix when placing it in a secondary fermenter, as extra air in the jug or carboy may cause souring.

Traditionally, t'ej is drunk about three weeks after it's made, but it's sometimes bottled and aged. You can also carefully pour or siphon it into flip-top bottles and put it in the refrigerator for a couple of days for a refreshing chilled, carbonated drink to go along with a spicy meal or to cool off with on a hot summer day. If you're going to wait for more than a couple of days, be sure to burp the bottles occasionally to avoid refrigerator grenades. As with any mead, I suggest saving a bit of the actively fermenting t'ej as a starter for your next batch (assuming you like how it turned out).

This is simple, traditional t'ej. There's no need to go any further than this, but if you're feeling adventurous, read on for some variations and funky ingredients that I've personally tried, and others I've read about.

"The Lost Vikings" Banana-Coffee T'ej

When I first started experimenting with wild-fermented t'ej, I learned that it could be made with ingredients such as herbs, berries, bananas, and coffee. The latter two intrigued me, as one of the earliest (and still one of the best) beers I made was a cherry-coffee-chocolate stout. The strength and bitterness of the coffee and hops balanced nicely with the sweetness of the cherries and malt. This combination also apparently had an additional effect.

Being a devout coffee drinker, I never noticed any caffeine buzz from drinking the stout. One morning, after I'd had some friends over the night before, I drank one with breakfast, joking that I liked to start out my day with coffee and beer all at once. My houseguests looked at each other and said, "You never told us that had coffee in it! No wonder we were up until three in the morning with the shakes!" I'm quite sure

I'd told them, but due to their proclivity for other intoxicants, they may very well have forgotten.

Using coffee in mead has a very different effect flavor-wise than it does in beer, as mead is like a delicate flower compared with the robustness of a stout. Also, bananas rot quickly, which means they will present some off flavors for the first year or so after fermenting; these will mellow after a couple of years in the bottle. I wasn't a huge fan of the banana flavor when tasting the initial fermentation, but I've enjoyed bottles that I let age for a couple of years. Yours may turn out different depending on how you experiment (t'ej is ripe for experimentation), but the ones I made had a bit of a musty flavor—which isn't necessarily a bad thing. An attendee of one of my workshops once told me it tasted like cellar (this could have been a result of a particular strain of bacteria that made it into my wild ferment as opposed to the recipe itself). Regardless, this is one of those meads that has an ancient feel to it and is best enjoyed out of a mead horn around a fire. I named it "The Lost Vikings" because I pictured a group of Vikings who had wandered a bit too far and ended up in the land of bananas, coffee, gesho, and honey.

INGREDIENTS FOR 1 GALLON (4 LITERS)

3 cups (700 mL) honey

12 cups (3 L) water

½ cup (120 mL) coarsely ground coffee beans

3–4 bananas

½ cup (120 mL) freshly brewed high-quality coffee (optional)

10–12 raisins

1–2 small pieces oak bark, gesho sticks, or 1 tea bag for
    tannin (optional)

PROCESS

1. Mix the honey and water at room temperature (or heated to below 100° F [38° C]) in an open-mouth vessel.
2. Stir several times a day until fermentation commences.
3. Add the coarsely ground coffee once fermentation commences.
4. Add the bananas, peeled and sliced, along with some of the peel for its wild yeast and malic acid.
5. Add the raisins and tannin.

6. Add the freshly brewed coffee if desired.
7. Cover the vessel with a lid and continue stirring for about five days.
8. Check regularly and remove the bananas as soon as you begin to smell them strongly, or if they've started to turn brown.
9. Drink young or strain or siphon into a carboy for several months and then bottle to age.

Dr. Yeti's Ethiopian Fire Mead/T'ej

This was the first mead I made using gesho sticks from Ethiopia. Technically it's a t'ej, but technically t'ej is a mead, so call it what you want. The original recipe can be found on my article of the same name at Earthineer. Its name was based on a couple of factors. First, as you surely already know from the preface, I write for Earthineer.com under the pseudonym RedHeadedYeti and thus am known by many as "The Yeti." My friend Steve Cole is known as Stickboy on the site and created an illustration for my bottle label.

T'ej is traditionally made in hollowed-out gourds smoked over a fire,[3] so I built a fire and held my fermentation crock over it to give the vessel a heavy dose of smoke. It was midsummer and I had recently picked some cayenne and poblano peppers from my garden, so I smoked some over the fire and added them to the ferment. I figured since t'ej was often consumed with spicy foods, I would just go ahead and get the sweet and spice in one sip, with a nice smoky flavor throughout. And thus you have the story behind the name.

That initial batch was unanimously recognized by my friends (and by me) as one of the best meads I'd yet made. It's not that I hadn't made good meads before, but this mead had some real character to it. Personally, I feel that it was because I had developed a relationship with the mead as I was making it. I used barm from a batch of mead I'd started the previous month when doing a mead-making workshop at the Whippoorwill Festival near Berea, Kentucky. I then nurtured that starter until I was ready to make mead out of it, built a fire to call forth the *bryggjemann*, and used ingredients from the garden I had toiled over as the primary flavoring. If you make this, I encourage you to go about it organically and impart your own spirit and homegrown ingredients into it.

INGREDIENTS FOR 5 GALLONS (20 LITERS)

Smoke—if you smoke the vessel and peppers over a fire, use wood you would use for cooking (hickory, applewood, cedar, etc.)

12 pounds (6 kg) light honey, such as wildflower or clover; the counterbalance of peppers and bittering agent make this mead ideal as a semi-sweet or dessert mead (add another 2–4 pounds [1–2 kg] honey)

Enough water to fill a 5-gallon (20-L) fermenter to the neck

25–30 organic raisins

A label for Dr. Yeti's Ethiopian Fire Mead designed by the author. Illustration by Steven Cole.

A handful of dried gesho or woody hops sticks and leaves; *or* oak
or walnut twigs, bark, and leaves

4–10 smoked or grilled cayenne, poblano, or other peppers of
your choice

Wild yeast, barm, or sweet wine yeast

PROCESS

1. Build a fire and smoke your fermentation pot beforehand.
A few drops of Liquid Smoke added to the must would
work as well. Other options are to smoke some wood chips
separately and let them soak in the mead for the initial fermentation. My editor recommended a smoked tea, Lapsang
souchong, as another option. I haven't yet tried this, but
black tea is great for tannins, so I can't see why not.
2. Mix the honey and water to the desired proportions in your
fermentation vessel.
3. Stir for several minutes with your totem stick.

A batch of Dr. Yeti's Ethiopian Fire Mead (t'ej) that has recently begun
spontaneously fermenting thanks to the yeasts on the gesho sticks and
leaves from Ethiopia, and lots of stirring with a totem stick.

4. Add the raisins, gesho, peppers, bittering herbs, or whatever else you desire.
5. Add a sweet wine yeast, barm, or wild ferment.
6. Ferment, rack, bottle, age, and drink.

Appalachian Viking T'ej

My friends Taye Spink and Camille Hyberger provided me with the honey for this recipe. At the time, they lived near me in Berea, Kentucky, and I brewed it using some gesho at the mead hall my brother built in the back 40 of our childhood Kentucky farm. Hence the name. Taye and Camille kept bees behind a Berea restaurant, in a very small yard—an example of how bees can be kept practically anywhere. The previous winter had been particularly brutal and they had lost their hives like most beekeepers in our area. But this loss was a boon to a mead maker. I made a couple of batches of mead that summer from honey and comb

Placing jars of honey and sorghum in a pot of water on medium-low heat can help dissolve the thick syrup to aid in working everything out of the bottle.

taken from hives that had died off. Taye and Camille's honey was likely clover-fed, but you can use whatever honey is available. I recommend a darker honey, since this recipe also calls for sorghum (or molasses), which will likely outperform a more delicate honey. In the end, use what's available.

INGREDIENTS FOR 5 GALLONS (20 LITERS)

3 gallons (12 L) water (give or take)

½ gallon (2 L) (6 pounds [3 kg]) honey with comb

Another ½ gallon (2 L) of a different type of honey, plus a pint or two more if you want a semi-sweet or sweet mead

2–3 ounces (57–85 g) gesho, oak bark, or other tannin

25 ounces (708 g) sorghum (or molasses)

Juice of 1–2 lemons or other citrus to taste (if you feel it is needed upon racking)

20–30 raisins

PROCESS

1. This is one of the few mead recipes for which I will recommend heating the must. If you create a tannin tea in advance, this isn't as necessary. However, since this is a t'ej, which is traditionally smoked, I like to make it outside over a low-heat wood fire. Also, a heated must will help break down the honeycomb, imparting its flavors into the mead.

2. Stoke a wood fire. You want it hot enough to fully infuse the ingredients and dissolve the honey, but not so hot that it will bring the must to a boil. Add the water, honey, tannin, and bittering agents to a pot large enough to hold everything without risking splashing, as you will be doing a lot of stirring.

3. The medieval recipes I based this on called for boiling the must for several hours. However, this not only kills off wild yeasts and good nutrients, but rapidly dissipates the must. Stir the must regularly while heating for about an hour, being sure to scrape the bottom with each stir to keep anything from sticking and burning.

4. Take the pot off the fire, set it aside, and place a lid on it. Let it sit for several hours or overnight.
5. In a day or two, pour the must into an open fermenter along with the sorghum and raisins. Proceed with fermentation by your desired method.

Not all mead must needs to be heated, but when you're making a whole-hive mead, it helps to warm the must to extract the maximum amount of flavoring from the comb. Soak wood chips in water beforehand and place them in the fire to impart a smoky flavor. Photo courtesy of Colleen Casey.

6. Once you've had an active fermentation for a couple of days, transfer the mead to a carboy. In addition to the standard equipment, you will need some sort of strainer to filter out the excess honeycomb and bits of other ingredients. You can use a brewing funnel with a strainer attachment, clean cheesecloth or other piece of cloth, or do it the Viking way as I did. Since the Norse and other ancient cultures (and a lot of modern cultures that still brew the traditional way) used plant matter such as juniper bushes or straw for filters, I elected to thoroughly hose down some garden straw, set it out in the sun to dry, and then place it in my brewing funnel. This caught most of the larger bits; the rest floated to the bottom during fermentation. You can use any edible, trustworthy plant matter.

7. Add enough water (or honey-water mix if you feel it needs to be sweeter) to fill the carboy to the neck, put in a cork and airlock, and wait. It will likely need at least one more racking, as this mead produces a particularly large amount of residue.

CHAPTER SEVEN

Herbal, Vegetable, Floral, Fruit, and Cooking Meads

There are holes inside all of us. Holes that can only be filled by certain plants. Empty space needing tree or stone or bear. Emptiness that can only be filled by some of the other life on this Earth. Other life with which we have evolved through a million years of coevolution. Without filling them we live a half-life, never becoming fully human, never being healed or whole or completely who we are. Never becoming completely sane.

—*Stephen Harrod Buhner,* The Lost Language of Plants[1]

Mead and wine can be made from practically anything edible. Country wines—and by extension meads—are a traditional way to preserve excess produce from the garden and orchard. Fermentation rituals are also a long-honored tradition for enhancing the healing power of plants and for building deep connections with "plant allies."

Vegetable Mead

Nearly any vegetable can be fermented into a flavorful mead or wine. The recipes I provide here can be used as bases for other meads with similar types and textures of vegetables. Also, if you don't have much honey available, it is perfectly acceptable to substitute some or all of the honey with sugar to make wine. People often turn up their nose when

I mention making mead or wine out of vegetables they would normally have no problem eating. However, you can make wine that will stand up to the finest of grape wines (and in many cases taste very much like a grape wine) from nearly any vegetable. The same goes with mead.

If you want the flavor of any fruit or vegetable to stand out in a mead or wine, it's best to save a portion of it for secondary fermentation. During primary fermentation, most of the flavor will dissipate as fermentation works its magic and creates alcohol. Don't let the thought of a certain vegetable's flavor turn you off from making mead or wine out of it. Alternatively, some vegetables, like alliums (onions, garlic, leeks, and the like), may impart too strong a flavor for drinking, but will work great as a cooking mead. You can even intentionally keep these meads out extra-long in open fermentation to turn them into cooking vinegar. That being said, one mead maker, Claude Denn, told me on Earthineer that he adds lots of extra garlic if he wants to make a cooking mead, or a bit less for a dessert mead (which has won him at least one award in the Mazer Cup mead competition). You're a Viking—don't let the naysayers stop you from having a bit of fun!

Mushroom and Garlic Mead Madness

Mushrooms as an ingredient in mead? As you surely know by now, I don't limit myself when it comes to ingredients. So long as you exercise caution and follow the proper protocols for growing, harvesting, and preparing mushrooms, they can be some of the most flavorful (and healthy) ingredients you can add to a mead, wine, or beer.

While there is a historical basis for making mead and beer with psychotropic mushrooms (which the Vikings almost certainly used), we're not going to touch on that here. Regardless of your thoughts on mind-altering, naturally harvested drugs, you should always be very careful about using any type of mushroom in alcohol fermentation other than those you have identified as edible with 100 percent certainty. If you're hesitant about using verifiably safe mushrooms in mead, consider that yeast itself is a type of fungus, so every alcoholic beverage you consume contains a colony of mind-altering fungi!

As part of my intent to grow or harvest from the wild all the ingredients I use for cooking and brewing, I grow mushrooms on logs and in mulch. The ones I don't eat fresh I dehydrate. When I first decided to use garlic in a mead, I had recently rehydrated some dried shiitake mushrooms that I harvested from one of the logs in my "mushroom forest" the prior summer and thought they would be a good addition to the garlic mead. To rehydrate them, I placed them in water for a few hours, which resulted in a tasty mushroom broth. You can also soak dried or fresh mushrooms beforehand in vodka or grain alcohol if you wish to "sanitize" them while drawing out the flavor. Take note, though, that many mushrooms should be cooked before eating or using for brewing. This is not only because they are difficult to digest otherwise, but also because cooking brings out additional medicinal qualities. If you're brewing a bragot (see chapter 8), or choose to heat the must for any other type of mead, simply chop up your mushrooms and toss them in near the start of the boil. Otherwise, make them into a tea that you can add, once filtered and cooled, to the secondary ferment or at the end of the boil.

INGREDIENTS FOR 1 GALLON (4 LITERS)
2–4 pounds (1–2 kg) honey
1 gallon (4 L) water
Yeast of your choice (I prefer to wild-ferment this one)
4–12 heads garlic (depending on how brave you are)
2 cups (475 mL) chopped oyster or shiitake mushrooms;
    or 1 cup broth or tea
10–12 raisins

The early stages of a garlic-shiitake mead.

PROCESS

1. The process isn't that much different than for any fruit, vegetable, or spiced mead. You can roast the garlic beforehand, but try not to get any cooking oil in the mead. I roughly chop or smash the garlic to get the juices flowing and put it directly in the must. For a stronger garlic flavor, put most of the garlic into the mead after active fermentation dies down. If you'll be using the mead for cooking or want people to keep their distance from you, put a bit of raw garlic in each bottle.

2. Add the mushroom broth at any time or leave it out. Wild-ferment, use a sweet or dry wine yeast, or add barm or backslop and proceed as with any other mead.

 Squashed Pumpkin Mead

You can use the following process and ingredients for any type of squash or other dense, low-sugar vegetable. I often end up with more squash than I know what to do with, which was the inspiration for making this cheap, easy mead. The key is to taste throughout and adjust. Squash doesn't have a lot of sugars, which can result in a dull mead that doesn't ferment well if you don't add the proper amount of honey (or sugar). When I first made this, it was tasty and sweet right after I cooked the must, but not so tasty after it had fermented. So when racking, I added honey to taste—and not only did the flavor improve, but it brought out more of the squash flavor. I like to drink some within a week or two of the initial fermentation, add some to a new ferment as barm, and save a few bottles to age.

INGREDIENTS FOR 1 GALLON (4 LITERS)

5–6 medium-sized butternut squash, pumpkin, or a mix of squash varieties

1 gallon (4 L) water

1 package or 2 teaspoons (10 mL) baker's or brewer's yeast (I used brewer's); *or* 1 cup (240 mL) barm

1½ pounds (.75 kg) brown or cane sugar

The must of squashed pumpkin mead before the solids have been strained out.

10–12 whole allspice (cracked or coarsely ground)
2–3 cloves
1–2 whole nutmegs (cracked or coarsely ground)
Juice of 1 lemon or 2 limes
1–2 pounds (.5–1 kg) honey

PROCESS

1. Bake the squash at 350° F (177° C) for 20 to 30 minutes. I recommend quartering and de-seeding them first. Check the squash regularly. When the skin starts to brown and crinkle and the flesh is soft, turn the heat off. When the squash is cool enough, peel off all the skin and remove any remaining seeds or blemishes.
2. Put the squash in a large stockpot and add water. Add all the ingredients (except the honey), and stir until you have a thick soup. Bring to a light boil and cook for one hour.
3. Remove the stockpot from the heat, put on a lid, and let the must cool to room temperature.
4. You have a few options for your initial fermentation. I tend to pour everything into an open fermenter, add yeast

or barm, and aerate regularly as I would a wild ferment. If you're adding yeast, you can let the must settle and rack as much liquid as possible into a narrow-neck jug. However, this can get messy and you'll need to rack again soon due to all the sediment. I like to wait instead until fermentation commences and then siphon the liquid, pressing the solids with a spoon to ensure I get as much flavor as possible infused in the liquid. This is still a bit messy, but your next racking shouldn't have as much sediment.

5. Once you've racked into a secondary fermenter, place an airlock in it and let it sit in a warm, dark corner for about a week. This is when it's important to adjust sweetness. Add 1 to 2 cups (240–475 mL) of honey to a 1-gallon (4-L jug) along with a bit of water and stir or shake well. Siphon the wine or mead off the sediment and stir or shake again. If it still doesn't taste quite sweet enough, add more honey. Remember, the sweetness of the honey will dissipate if you leave it to age to a dry mead, so don't skimp. Taste it regularly, add honey when needed, and drink as much as you desire while it's young and bubbly.

6. This mead will need a fair amount of sweetness and some tannins and acids or it can turn out a bit bland. I recommend that you continue to tweak the flavor until you're happy before bottling.

Metheglin (Herbal and Spiced Mead)

We've already discussed several botanical ingredients to use as flavor enhancers when preparing mead, but what of the many other assets our little green (and brown, and yellow, and . . .) friends can provide us? Our ancestors didn't just put random plants in their brews because they liked the flavor. Usually there was a specific, known medicinal or stimulatory benefit of using each plant. While certainly a degree of trial and error was part of the process of learning what these plants could do for us, much of what they learned came from "speaking" to the plants themselves. Native Americans, Appalachian mountain folk, Norse healers and herbalists—and many other cultures and subgroups with a deep connection to the land—have long recognized that we don't look for a plant that meets our healing needs, but we listen for the plant to speak to us first.

Mead made with herbs and spices for intentional flavoring or medicinal qualities can properly be referred to as metheglin. Depending on the source—and time period—you may also see this spelled as *meddyglfyn*. The latter is a Welsh word from which the modern spelling likely originates. Its root, *meddy*, is shared by several Welsh words pertaining to medicine, healing, or contemplation.[2] Although herbal meads were certainly made by ancient cultures before *metheglin* became the standard descriptor, the use of herbs in mead was clearly associated with physical and mental healing by Northern European cultures due to its common use as a medicine or "healing liquor." It makes sense that honey and fermented beverages would have been used for the long-term storage of plants used for medicine, as both are known for their preservative qualities, and for their ability to enhance the effects of healing plants.

Take care not to just haphazardly add large amounts of herbs or spices when creating a metheglin. Many can have particularly strong flavors, even in small amounts, and can lead to an overly medicinal taste. Don't let that stop you from experimenting—just be sure to do your research and start out with small batches. If the resulting flavor isn't to your preference, note its medicinal qualities and save it as a healing tonic—or simply brew with the health benefits at the forefront of your intentions.

When added to mead, some herbs—yarrow and wormwood in particular—were historically noted for producing particularly nasty hangovers if drunk in excess. This seems to have been a point of pride with farmhouse brewers who competed with their neighbors to produce the strongest ale. The flavoring, intensity of the hangover, and "madness" caused while drinking were often contributing factors to the farmer's pride in his or her particular brew.[3] Herbs such as bog myrtle and wormwood are potentially harmful to pregnant women. Other herbs and plants to avoid during pregnancy include barberry, black cohosh, cayenne, comfrey, ginger, goldenseal, mugwort, Oregon grape, and yarrow.[4] Always research and, if necessary, contact a trusted medical or herbal expert before using any herbs to excess in brewing.

The Mad Monk's Medieval Metheglin

This is a whole-hive metheglin-style mead, meaning that I made it using both honey and the remnants of the hive (that is, the comb) that came with it. I've made variations with a similar herbal combination over the years, but the recipe below made a particularly good mead, and I think the honey and honeycomb played a large part in it. My friend Matt Wilson provided me with some of his top-bar hives for this batch, which contained a particularly delicious honey with a mild apricot flavor. I had dug up some medieval mead recipes and decided to follow them as closely as possible while avoiding techniques I know to be unnecessary, such as boiling the must. I must say, this turned out to be an incredibly flavorful semi-sweet mead, which I've enjoyed both young and aged.

INGREDIENTS FOR 5 GALLONS (20 LITERS)
1 sprig dried rosemary
2 teaspoons (10 mL) dried sage
4–5 teaspoons (20–25 mL) dried nettle, red clover, and
    raspberry leaf

When preparing a whole-hive mead, extract what honey you can first and then submerge the comb (dead bees, propolis, and all) in water overnight before adding it to the must.

2 teaspoons (10 mL) grated fresh ginger

A bit of oak bark or a couple of oak leaves

4–5 gallons (16–20 L) good, clean water

12–15 pounds (6–7.5 kg) light to medium-amber honey and
(if you have it) comb

Wild yeast or barm

Juice of 1–2 small lemons, and additional citric acid upon
racking, if needed

20–30 organic raisins

PROCESS

1. Make a tea out of the herbs and oak bark. It's best to do this part a day or two before mead-making day; simply steep the herbs in hot (not boiling) water the night before to get the optimal health and flavor benefits.
2. Submerge as much of the hive as you can under water into a gallon jar or other container.
3. Prepare a smoky, low-heat fire, and place all ingredients except the yeast, lemons, and raisins in a pot; set the pot on a rack over the coals of the fire or over some heated, flat rocks.
4. Stir the must and skim off any wax or other scum that floats to the surface.
5. Take the pot off of the fire after 30 to 45 minutes, cover with a lid, and set aside for several hours or overnight.
6. When the mead is "blood warm" (warm to the touch or room temperature), add the yeast, backslop, or barm, or proceed with wild fermentation.
7. Add the lemon juice and raisins.
8. Age, rack, drink, and bottle.

Chai Metheglin

The spices used for chai tea make a wonderful combination for mead, and many of them show up in medieval recipes, so why not combine them all in one batch? Spices, particularly those used for chai tea, were in high demand during the Middle Ages, as the development of spice routes from Asia provided new avenues for people—royalty in particular—to impart exotic new flavors to their normally bland food and drink. You can use chai tea bags (one per 5-gallon [20-L] batch of mead), but you have much more control over the final flavor if you gather the individual spices. Be warned that some of these spices—especially clove, fennel, and cardamom—can easily overpower the rest of the flavors. I suggest making a tea beforehand. If you like the flavor, use it for your mead; if you feel it needs more or less of a certain spice, drink the tea and make another batch for your mead.

INGREDIENTS FOR 5 GALLONS (20 LITERS)

3–5 cloves
4–5 teaspoons (20–25 mL) crushed nutmeg
4–5 teaspoons (20–25 mL) crushed cardamom
3–4 cinnamon sticks
2–3 teaspoons (10–15 mL) fennel
2–3 teaspoons (10–15 mL) crushed black peppercorns
2–3 crushed star anise pods
4–5 teaspoons (20–25 mL) grated or crushed fresh ginger
4–5 gallons (16–20 L) good, clean water
12–15 pounds (6–7.5 kg) medium-amber to dark honey
Yeast (sweet wine, wild, or otherwise)
20–30 organic raisins

PROCESS

1. Make a tea out of the spices the day before and let it cool.
2. Blend the honey and water in a pot on the stove, taking care not to let it come to a boil.
3. Add the tea and stir well.
4. Pour into an open fermenter and proceed with wild fermentation, or pour or siphon into a carboy, adding yeast or barm about halfway through the process to ensure proper aeration.
5. Rack after a couple of months and taste. If you feel a stronger chai flavor is needed, add another cup or two of chai tea, leaving out any spices you feel are already represented well. You can do this all the way up to bottling, but if you are satisfied with the flavor at any point, leave it be.

The must of a freshly prepared chai metheglin.

6. Wait until fermentation is complete—likely close to a year after making it—bottle, and wait at least six months before sampling.

Odin's Golden Tears of Honey Joy Cinnamon-Vanilla Metheglin

This recipe is based on one that my friends Josh Parker and Dave Brown came up with. I say *based on* because, although they made similar meads with astounding degrees of success in the past, they allowed themselves to get too involved in the technical details with each subsequent brewing session (likely from reading too many modern mead recipes).

After they made a batch or two that didn't turn out as good as they'd hoped, they decided to pull out all the stops. Rather than going back to the drawing board and figuring out where they could simplify, they went to great effort to add nutrients and acid blends purchased from a homebrew store on a strict schedule, take hydrometer and pH readings religiously, and sanitize excessively. Now, I realize this may just be part of the process for some brewers, but after they went to all that work and still weren't pleased with the results, they were ready to give up and didn't sound like they were enjoying themselves anymore. They dumped out some of the bottles and passed the rest along to me to hopefully mix with a new batch and "fix." After tasting a bottle that had aged a couple of years, I noticed that, while still not ideal, it was much less flat and bland than early samples. I have high hopes for it. I pass along this story merely to note that, while you can get as complicated as you want, the best policy is to shoot for simplicity. Also, never toss a mead, no matter how bad you think it is! Sometimes time, or other palates, can work wonders. The following recipe is a stripped-down version of Josh and Dave's original recipe that Dave and I came up with.

INGREDIENTS FOR 5 GALLONS (20 LITERS)

3–4 cinnamon sticks
2–3 teaspoons coarsely ground whole allspice
12–15 pounds (6–7.5 kg) medium-amber to dark honey
4–5 gallons (16–20 L) good, clean water
2 vanilla beans
Yeast (sweet wine, wild, or otherwise)
3 bags vanilla-rum tea (optional)
20–30 organic raisins

PROCESS

1. Make a tea out of the cinnamon and allspice the day before and let it cool.
2. Blend the honey and water in a pot on the stove on low heat, or add straight to the fermenter with no heat.
3. Twist one vanilla bean with your hands to bring out the juices, or slice it down the middle, and add it to the must.
4. Add the yeast/barm or wild-ferment.
5. When the initial fermentation subsides, add another vanilla bean, the spice tea, and 1 to 2 cups (240–470 mL) of vanilla-rum tea if desired.
6. Rack two or three times over six months to a year, adding more vanilla or spices as needed. Sometimes a drop or two of vanilla extract when racking will help bring the vanilla flavor out if you don't feel enough of the vanilla profile is present.
7. Bottle, age, and experience the joy of Odin's tears.

Peach Horehound Mead, a Respiratory and Digestive Tonic

White horehound is an annual and is considered an invasive species in some areas due to its tendency to take hold and spread in even the barest of soil. The respiratory and digestive benefits of horehound (*Marrubium vulgare*) are well documented in folk medicine. Its bittering agents serve as an expectorant, helping to loosen mucus and assuage upset stomachs by stimulating stomach secretions. It was also traditionally used to expel poisons and as an "anti-magical" herb. Take care to not confuse white horehound with black horehound (*Ballota nigra*), a foul-smelling but edible variant that is also known as "black stinking horehound."[5]

Although you can make this as a straight herbal-only metheglin, I like to add seasonal fresh fruit to it. I once added horehound from my garden in Kentucky to mead that I made from South Carolina peaches, North Carolina sourwood honey, and water from a North Carolina spring. The combination of horehound's powerful health benefits

A batch of Peach Horehound Mead in the works.

and the antioxidant properties of fresh peaches make for a great tonic for fighting off winter illnesses. Other herbs you can add to boost the effects are: hyssop, rue, licorice root, and marshmallow root. To top it off, although horehound is considered a bitter herb, the bittering effects are mild (some people seem to consider it more bitter than I do) and balance with the sweetness of the peaches and honey to make for a pleasant drinking experience. It does have a strong flavor, though, so keep in mind that a little goes a long way.

INGREDIENTS FOR 1 GALLON (4 LITERS)
2 pounds (1 kg) light- to medium-amber honey with comb
1 gallon (4 L) good, local spring water
Wild yeast or barm
3–4 average-sized peaches, berries, or other sweet fruit
1 sprig fresh horehound or 1–2 ounces (28–57 g) dried
½ ounce (14 g) each hyssop, rue, licorice root, or marshmallow
    root (optional)
10–12 organic raisins
½ cup (120 mL) orange juice

Fruit-Based and Flower Meads

A melomel is simply a fruit-based mead. But the revered Namers of Mead Styles apparently don't like to keep things simple. If you make a mead with apples, you have cyser, which is similar to cider. A pyment is a grape-based wine that uses honey for its fermentable sugar or, if you prefer, a grape mead. Since the technique is similar for nearly any type of melomel, I'll provide the basic process and a couple of recipes and variations. Summer is a great time to start preparing for fruit-based meads, as the profusion of fresh fruit available provides an abundance of options. If you don't have the time to make mead during the busy harvest season, you can freeze some fruits and store others in a dark, cool area until you're ready to use them.

For a standard melomel, simply follow the basic wildcrafted mead preparation guidelines in chapter 5. You can add most fruit whole to the must, but for larger fruits you'll need to chop and remove the skin (if it's

JESSE FROST
THE SIMPLICITY OF NATURAL MEAD AND WINE MAKING

Jesse Frost is a writer and organic farmer who lives in western Kentucky with his wife, Hannah. His method of making mead and wine echoes the simple, natural methods I have outlined in this book. In his book *Bringing Wine Home* (a planned three-part series), he muses on his path to eliminating unhealthy foods and addiction to alcohol and tobacco from his diet to living healthily off the land while enjoying natural, homemade mead and wine in moderation. His journey takes him from a childhood spent in Colorado and Kentucky, to working in a wine shop in Brooklyn, to traveling to France to discuss biodynamic farming and winemaking with vintner Bertrand Gautherot, to returning to Kentucky to try his hand at organic farming.

His article on making wild-fermented mead shows just how simple it can be to make mead naturally. His process boils down to this: Mix honey and water, add whatever fruits or herbs strike your fancy, stir, let ferment, age and bottle if you desire, and drink. Kind of makes you wonder why an entire book would need to be written on this simple but nearly lost art. I have enjoyed speaking with Jesse and attending his workshops on making what the wine industry likes to call "natural" wines, which he describes as "unrefined, unfiltered, wildly vibrant wines—volatile and funky."[6]

Author Jesse Frost demonstrates natural winemaking during WildFest 2014 at Cedar Creek Vineyards in Somerset, Kentucky.

bitter or inedible). It is generally not necessary to juice the fruit first, as the fermentation process works wonders for getting to its essence.

You can add as much fruit as you'd like to these recipes without worry. Unlike spices and some botanicals, excess fruit doesn't risk overpowering the mead (at least not in an unpleasant way). The more fruit you add, the less honey you'll need, as fruit provides plenty of its own sugar. When I asked wild-fermentation expert Sandor Katz about his preferred ratios for fruit meads, he told me that he generally does about a 1:8 honey-to-water ratio for meads that are essentially a bunch of fruit covered with a honey-water mix. For "a substantial amount of fruit but not that much," he goes with a 1:6 ratio. For all his meads, he starts with a 1:4 ratio when working with just honey and water and modifies from there based on what other sugars he adds. Keep in mind that many fruit meads won't taste much like the fruit they're made from once fully fermented if you only add fruit with the initial fermentation. If you want a stronger fruit flavor, add more after primary fermentation has commenced, and continue to add with each racking until you're satisfied. This isn't to say you need to make an overpowering fruit mead. Some people, myself included, prefer most of their melomels to have just a touch of fruit flavor with each sip.

 ## Berry Melomel

Almost any kind of mild to sweet berry can be used to make this mead. Feel free to combine berries for a berry-medley melomel. If you add tart berries to the mix, take care to not overdo it unless you're the sort who likes a good pucker. Berry meads are well suited for wild fermentation, as the plethora of yeast on the berries and in the honey inevitably makes for a vigorous fermentation. You can also finish off melomels with a commercial yeast. See table 4.2 (on page 65) for suggestions on yeast strains that are good for melomels. Elderberries make for particularly flavorful berry melomels.

INGREDIENTS AND PROCESS FOR 5 GALLONS (20 LITERS)

1. Pick and de-stem as many berries as you can. The best way to de-stem small berries like elderberries is to freeze them and work while they're frozen. Try to get all the stems. It

won't hurt if a few make it in, but too many can cause an excess of tannins. Place the berries in a bowl with enough water to cover them. Any overripe berries will float, along with leaves, insects, and other crud.

2. Remove unwanted materials with a strainer, pour off the water, and add as many berries as you desire to a 5- to 6-gallon (20- to 24-L) bucket or crock, saving a pound or two for secondary fermentation if you want a strong berry flavor. Alternatively, if you've been extra careful when de-stemming, go ahead and dump everything in a pot of warm water, skim off any berries that float, mix in your desired amount of honey (see below), and proceed.

3. If you fill the fermenter three-quarters full or more, prepare a 1:8 honey-water mix and pour it over the berries until they're just covered. If you only fill it a quarter to half full, go with a 1:6 mix. Use a light to medium honey to draw out the flavor of the berries.

4. Most fruits will contain some degree of acid, but it doesn't hurt to add a cup or two of orange juice. Inevitably, elderberries will provide a fair amount of tannin, so don't feel

When you're making elderberry mead, pour stemmed berries into a bucket or pot of water, and skim off any that float to the top.

you should add extra. Adding 15 to 20 raisins for wild yeast and nutrients isn't a bad idea, though.

5. Stir and lightly mash the berries into the must and cover the fermenter with cheesecloth. Continue to stir whenever you get a chance, but most fruit meads tend to take off pretty quickly with minimal stirring.

6. Be sure to place the fermenter in a container that will catch any overflow, or be prepared to clean up a mess. You can also put this straight into a carboy, but don't fill it any more than three-quarters until it's had at least a week to finish fermenting. It's also a good idea to use a blow-off tube rather than an airlock. To do this, take the stopper, put a siphoning tube through the hole, and submerge the other end of the tube in some water.

7. Two or three weeks after initiating fermentation, rack into a carboy, leaving most of the berries behind, and add as many more berries as you wish for flavoring, keeping in mind that they will cause additional fermentation.

This elderberry melomel required a blow-off tube and room in the carboy for initial fermentation. Once it died down, the author added more honey-water, let it age for a couple of months with the berries, and then racked it off the berries and lees.

8. Rack every couple of months until the mead is the color of a nice, clear juice and bottle when ready. Berry meads are great candidates for bottling early (within six months) in thick-glass bottles for a carbonated fruit mead.

Cabin Fever Cyser

At its core, cyser is simply fermented cider and honey. To the Norse, fruit was a symbol of the bounty of short summers as well as of immortal youth and beauty. The Norse goddess Idun was lighthearted and full of beauty. She personified the continual rebirth of spring and was therefore always beautiful and young. Fittingly, she was courted by and eventually married Odin's son Bragi, who personified music and poetry, and was also eternally young. Bragi was often called upon by skalds, who were sometimes known as *Braga-men* or *Braga-women*. His name was toasted at festive occasions such as Yuletide celebrations or funeral feasts, accompanied by a vow to accomplish some great deed in the upcoming year. Hence, his name came to be the root word of verb *brag*.[7] Idun's fruits, represented by apples in most versions of the myth after their introduction in the Middle Ages, were precious to the Norse, and she was reputed to greet the gods in Asgard daily with a basket of apples that bestowed immortal youth and a touch of beauty to all who ate them.[8]

Up through the early 20th century, cider (and to a degree, cyser) was a common farmhouse product due to the abundance of apples in agrarian societies. Even through the 1970s, the most common way to make cider or cyser was through wild fermentation. Apples from orchards that don't use pesticides teem with wild yeast and bacteria, much of it favorable to producing complex flavors. While any type of apple juice can be fermented into cyser, some apples impart much better flavor than others. Generally you should seek to blend two or more varieties of sweet and tart apples, depending on what type of flavor profile you're seeking. The quality and ripeness of the apples you pick, along with the honey you use, will be the ultimate deciding factor. In *The Cheeses and Wines of England and France*, author John Ehle sums up his method for selecting apples for cider: "The sharp class of apples contains a high percent of total acid, but almost no tannin . . .

The . . . sweet class of apple normally has less acid . . . but more tannin. Finally, the bittersweet apples contain a great deal of tannin, normally exceeding .2 percent. A mixture of these apples can yield a cider balanced in acid (about .6 percent), and tannin (.10 to .15)."[9] Of course, when making cyser, take note that you will have an excess of sweetness to consider. Use the ratios of honey to water referenced in the Show Meads section of chapter 6 to determine what level of sweetness you want first, and then determine your preferred acid-to-tannin ratio using the formula above.

As with other types of mead, don't fret too much. Work with what you can obtain and don't overthink it. Claude Jolicoeur, author of *The New Cider Maker's Handbook* and a mechanical engineer by profession, suggests the "Keep it simple, stupid" (KISS) engineering-design principle, or as ethnobotanist and mead maker Marc Williams likes to put it, "Keep it simple, sweetheart." Following are the core tips (I just can't get enough of those *core* puns) you should consider in gathering and selecting apples:

- Pick as many as possible, at least three bushels for a 5-gallon (20-L) batch.
- Choose apples from small, traditional orchards rather than commercial operations.
- If you grow your own apples or talk to the arborist or property owner of the orchard you pick from, select older trees and those that have been minimally fertilized (too much nitrogen leads to increased water and thus decreased juice).
- Don't just go for the "perfect" apples; scabs, evidence of insects or worms, or a bit of rot are common in naturally grown fruit (just be sure to take the time to cut off any bad parts and remove any worms before juicing).
- Look for late-maturing apple varieties and pick near the end of the season.
- Leave apples to "sweat" in a cool, dry place for several weeks before juicing; when they're soft to the touch, they're ideal for cyser.

Although these are all tips I agree with, I should note that this list was paraphrased from a more comprehensive list in Jolicoeur's book.[10]

I don't always have time to juice apples during the hectic harvest season, so one thing I'll often do is store my apples in my deep freezer in trash bags. This also softens the apples up when you thaw them, making for easier pressing or juicing. If you don't have access to a press, you can use a kitchen juicer, but be sure it is a high-powered one. My Acme Supreme Juicerator works well, but does require that I change the filter between every quart or two of juice. The alternative to these options is to simply purchase apple cider (not juice) to ferment. I'll often buy 1-gallon batches and ferment straight in the jug. Cider with no preservatives, particularly when mixed with raw honey, provides an ideal environment for wild fermentation (even in a narrow-neck jug with the cap off). It can't get quicker and easier than that. John Ehle sums up the simplicity of making cider naturally: "In the old days in England, the ciders were made on farms, just as were cheeses . . . The fruit was not washed. The yeasts were on the peels, after all. And the juice was not sulfited. The natural yeasts and bacteria were left to do their work. The English makers thought to be the best of that time also prided themselves on not adding water or sugar. They simply crushed the fruit and let the yeast convert the sugar into alcohol. That was all."[11] I would add to his statement that excess honey or brown sugar is helpful to add if you want some residual sweetness in cyser while still keeping to the simplicity of farmhouse brewing.

INGREDIENTS FOR 5 GALLONS (20 LITERS)

2 bushels of a quality, late-season, high-sugar, medium-acid apple variety

1 bushel of one or more additional varieties of apples

6–8 pounds (3–4 kg) honey

1 pound (.5 kg) dark brown sugar (optional)

1 gallon (4 L) water

1 cup (240 mL) raisins

1 teaspoon (5 mL) yeast nutrient (another cup of raisins should also work, but keep in mind that cyser requires a lot of nutrients)

1 5-gram packet Lalvin ICV D-47 yeast (or a champagne yeast for a dry cyser), or wild yeast

PROCESS

1. Prepare at least 4 gallons (16 L) of apple juice.
2. Add the juice to a wide-mouth or narrow-neck fermentation vessel (open fermentation isn't required due to the strong presence of wild yeasts on the apples).
3. Mix the honey, sugar, and water, and blend it with the juice.
4. Add the raisins and nutrient and stir well to aerate.
5. Sprinkle the yeast on the surface of the must and stir again, or proceed with wild fermentation.

OPTIONS AND RACKING SUGGESTIONS

For the most part, you can proceed from here as with any other mead. Whether you wild ferment or not, it is a good idea to leave 4 to 6 inches (10–15 cm) of head space in your initial fermenter, as cyser tends to ferment vigorously. It also produces a lot of sediment, so you'll need to rack a few times. Once the initial fermentation subsides and you've racked for the first time, fill the head space with more cider or with a honey-water mix. Additional ingredients that you may consider are: tart fruit such as cranberries or persimmons, and spices such as cinnamon, cloves, nutmeg, or cardamom.

Cysers and other fruit meads will often produce large amounts of krausen (foam) due to the multitude of yeast on the skins of the fruit.

If you don't have the time to juice apples and just want to make a quick, simple cyser, here's what to do:

1. Buy a gallon (4 L) of quality, natural cider and pour about a pint out to drink or save for racking.
2. Add a cup of honey mixed with water (you'll need to remove more cider to do this) for cyser, or leave the honey out and proceed with a normal cider.
3. Add 1 packet (5 g) of a sweet wine or champagne yeast, or simply leave to wild-ferment.
4. Add a handful of raisins.
5. Put the lid on loosely or cover with plastic wrap.
6. Rack to an airlocked vessel and continue to rack as needed until fermentation has subsided, usually about six months.
7. Bottle and age for one to two months.
8. At any point during this process, don't be afraid to taste it and, if you like it, drink it all up while sweet, young, and frothy.

Flower Mead

The process for making flower meads isn't that different from the process for melomel or other wild-fermented meads. You can also combine flower petals with fruits and a floral honey for a particularly divine mead that will be oozing with light, summery flavors.

INGREDIENTS AND PROCESS FOR 5 GALLONS (20 LITERS)

1. When the first flowers bloom in the spring, take every chance you can to start gathering flower petals. Be sure you pick flowers you know to be edible, and avoid picking from beside roads or areas that may have been treated with insecticides. Only use the petals; the stems and sepals (the green part in the center that holds pollen) can cause excessive bittering.
2. When you're ready to make some mead, set the flowers out the day before to thaw (see the "Flowers for Flower Mead"

FLOWERS FOR FLOWER MEAD

Edible flowers are ideal for wild-fermented mead due to the wild yeasts they contain. I rarely repeat specific recipes, as my flower meads are based on seasonal availability and the amounts I am able to harvest. Some flowers have very small petals, making it tedious to gather enough for a batch of mead in my busy homesteading life. Hence, I will often use several varieties and add fruit or other botanicals. A mead made with just flowers can be a divine thing, though.

When I pick small wildflowers such as violets or dandelions, I pick on a dry sunny day and place the petals in a brown paper bag, allowing them to dry as I take breaks to attend to other tasks. If I don't get quite enough for a mead, I store them in the freezer in tightly packed glass jars or plastic bags until I have enough. I sometimes employ my young daughter and her friends to help pick wildflowers and other "yard herbs," although this requires that I go through later and pick the good parts off. My dad learned this the hard way. He had us kids pick dandelions for a dandelion wine and didn't think to pick the greens out. To this day he says it was the worst wine he ever made.

Straining a flower mead made from wild violet, rose of Sharon (hibiscus), and dandelion.

Use the following list for tips on flowers to use. The measurements are the minimal suggested amounts per 1-gallon (4-L) batch. For any of these, you can substitute about ½ ounce (14 g) of dried petals per pint (475 mL) of fresh. For wild yeast, add a small handful of flowers during initial fermentation. To get the full flavor of the flower, wait until secondary fermentation. Add an ounce or two at a time and taste-test during rackings if you want to really fine-tune the flavor.

- Dandelion: 2–3 pints (950–1420 mL)
- Elderflowers: ½–1 pint (240–475 mL)
- Wild violets: 2–3 pints (950–1420 mL)
- Marigold: 1–2 quarts (1–2 L)
- Honeysuckle: 1–2 pints (475–950 mL)
- Roses: 2–3 ounces (57–85 g) (flavoring of varieties varies drastically)
- Gorse: 2–3 pints (950–1420 mL)
- Hawthorn (mayflower): 1–2 pints (475–950 mL)
- Rose of Sharon (hibiscus): 2–3 quarts (2–3 L)
- Lavender: 1–2 pints (475–950 mL)
- Purple clover: 2–3 pints (950–1420 mL)

    sidebar), laying them flat or placing them in a colander. Don't leave them for more than a day or they'll start to wilt and mold. You can also steep them in hot water, but don't boil the water if you want to make a wild ferment.

3. Mix a ratio of honey to water appropriate for a dry, semi-sweet, sweet, or sack/dessert mead, add fruits or other desired ingredients, and proceed as with any wild ferment. Be sure to add 1 cup of orange juice per gallon (4 L) or the juice of one lemon, as well as 10 to 12 raisins: Floral meads need just a bit of help to get the yeast working and fully extract the desired flavors.

CHAPTER EIGHT

Bragots, Herbal Honey Beers, Grogs, and Other Oddities

A s we've learned, the definition of *mead*—depending on who you talk to—can be as narrow as honey, water, and yeast with no other additions, and as broad as pretty much any fermented beverage that uses honey as a distinctive flavoring and fermentation agent. Since honey can be used to sweeten wine or to prime beer bottles for carbonation, I don't go as far as to say that *any* alcohol with honey in it is a type of mead, but neither am I strict about what makes a mead.

In the early history of alcohol, variations on the words that eventually came to be *mead, ale, cider, beer,* and *wine* would often be used interchangeably (see chapter 2). While I stick more closely to what those terms mean today, I won't limit my discussion here to beverages that might not be considered by purists to be mead. One such example is bragot (other name variations include *braggot, brag, bragio, brakkatt,* and *bracket*). This heavenly combination of malt, honey, and spices was standard for many centuries and, like more vinous meads, came in a dizzying array of varieties. The mass commercialization of brewing in the 19th century virtually wiped bragot from the consciousness of drinking society, but thankfully the homebrewing movement of the late 20th century is helping to bring it back.

The basic definition of a bragot is an alcohol made from honey and some sort of grain. It can be an ale fermented with honey added to the wort, an ale blended with an already fermented mead (or vice versa), or an ale brewed with honey and spices. Traditional Welsh ale (or Welsh bragawd) was a type of bragot, as Randy Mosher notes in his book *Radical Brewing: Recipes, Tales, and World-Altering Meditations in a Glass*:

> From ancient times the Welsh were famous for their honey beer, bragawd, and this lasted right up until the Industrial Revolution, when most of the rustic old-time brews faded away. By 1800, the recipes for Welsh ale include no honey. A Welsh ode to a drinking horn recounts in 1056:
>
> > Cup-bearer, when I want thee most,
> > With duteous patience mind my post,
> > Reach me the horn. I know its power
> > Acknowledged in the social hour:
> > Hirlas [name of the horn], thy contents to drain,
> > I feel a longing e'en to pain;
> > Pride of the feasts, profound and blue [referring
> >   to the silver from which it was made].
> > Of the ninths wave's azure hue,
> > The drink of heroes formed to hold,
> > Wih art enrich'd and lid of gold!
> > Fill it with bragawd to the brink,
> > Confidence inspiring drink.[1]

Basic Bragots

Be warned—if you're a beer lover, you may find yourself obsessing over bragots once you get a taste.

A Simple, Hoppy Extract Bragot

This is one of the simplest beers I've ever made, and even after years of beer and mead making one of the best. There's something to be said about going back to the basics, and this beer has all the essentials of both a fine beer and a fine mead. The pale malt, which is a common light malt for English ales, is accentuated by the amber's darker nuttiness, resulting in a full-bodied, malty flavor. The honey and malt combination lends a heavy dose of sweetness, but the bitterness of the hops counterbalances the sweetness, resulting in a satisfying experience with each glass. I highly recommend starting with this bragot before experimenting further.

INGREDIENTS

5 gallons (20 L) water

6 pounds (3 kg) amber dry malt extract

2 pounds (1 kg) pale dry malt extract

6–12 ounces (170–340 g) Cascade, US Fuggle, Sorachi Ace hops, a combination of two or three of these, or any other hops you feel like experimenting with (I often use only Cascade; if you're using multiple varieties, Cascade is a nice one to add at the end)

9–12 pounds (4.5–6 kg) light- to medium-bodied honey such as sourwood or wildflower

2 packets (10 g) Lavlin ICV D-47 wine yeast or Safale US-05 ale yeast, or 2 cups (475 mL) barm (keep in mind that bragots need very strong yeast, making wild fermentation unpredictable)

2 teaspoons (10 mL) each yeast nutrient and yeast energizer, or 2 cups (475 mL) raisins

PROCESS

1. Bring the water to a boil, then carefully, slowly add the extracts, stirring constantly to keep the wort from sticking to the bottom. Continue stirring and watch the heat until you've brought the wort back up to a boil.

2. Set a timer to 60 minutes, turn the temperature down, and keep stirring until you have a steady rolling (but not rollicking) boil. If you have brewed beer before, you know how important it is to watch a boiling kettle. Extract (and honey) can cause a monumental, sticky mess if allowed to boil over. Once you've had it happen once, you'll know to never let it happen again. When it does happen, take the pot quickly off the heat, stir like a berserker Viking, wipe the mess up with a wet towel, and proceed. Do *not* leave it until the next day to clean up. I did this with my first

brew in my Seattle apartment, which proceeded to harden into something akin to cement. Even after attempting to moisten it and chip at it with a hammer and chisel, I never really did get it all off.

3. Add the hops according to the following schedule (add a bit more to each stage if you're adding extra, keeping in mind that the earlier you add them, the more they'll contribute to bittering; the later, the more they'll contribute to aroma):

   • 60 minutes: 3 ounces (85 g)
   • 15–30 minutes: 2 ounces (57 g)

4. You can add the honey early in the boil, which will still result in a tasty bragot, but as with other kinds of mead I prefer to avoid boiling my honey, so I wait to add it after I've cut off the heat and stir it in. If you stir the honey in right after cutting the heat, you'll pasteurize it without killing off too much aroma. Wait a few minutes if you don't want to pasteurize the honey. This will result in an even stronger honey bouquet, and keep intact the honey's nutritional properties.

5. You'll want to cool your wort down as quickly as possible to get it to ideal yeast-pitching temperature, which is 60–70° F (15–21° C), or "blood warm" to the back of your hand. The standard methods for doing this are to set the brew pot in an ice bath in a bathtub or large sink, or to use a copper immersion wort chiller. Add yeast or barm.

6. If you'd like to dry-hop for extra hoppy aroma, add 1 ounce (28 g) of your preferred hops now.

7. Depending on whether you're doing open or closed fermentation, place on the fermenter a lid and airlock (for a bucket), cheesecloth (for a crock), or drilled stopper with an airlock (for a carboy), and set it in a warm, dark corner. You should see active fermentation within 24 hours. If not, give it a bit longer and see my tips on stuck fermentation in the Troubleshooting section at the end of this book.

8. Once it ferments, give it three to five days for the fermentation to slow down and then rack into a carboy, leaving

A wort chiller for quickly cooling heated beer or bragot wort or mead must to fermentation temperature can be made easily using materials from a hardware store for under $30.

as much of the trub (the beer version of lees) behind as possible. Add an airlock and wait a week or two.

9. You can now prime bottles as you would for a standard beer or sparkling mead (see page 101). My standard primer for a 5-gallon (20-L) batch is ¾ cup (180 mL) of corn sugar or ½ cup (120 mL) of honey dissolved in 2 cups (475 mL) of water. You can find priming calculators online that will give you the proper priming amounts for a wide range of sugars. They also allow you to plug in factors such as the style of beer you are bottling (if bragot isn't an option, barleywine is a good alternative), the amount, the desired volume of CO_2, and the temperature of the must. I like to use Northern Brewer Homebrew Supply's calculator (www.northernbrewer.com /priming-sugar-calculator). In addition to corn sugar and honey, other types of sugars they list as options are: sucrose, turbinado, corn syrup, brown sugar, molasses, maple syrup, sorghum syrup, DME (dry malt extract), Belgian candy syrup and sugar, black treacle, and rice solids. I haven't used

all of these, but sometimes I'll mix a bit of DME in with my honey. Whichever sugar source you decide on, add the sweetened water to the beer and stir it in slowly to prevent aeration that may add extra CO_2 to the bottle.

10. Bottle and wait two to four weeks before sampling. It should be fully carbonated at this point, and taste mighty fine, but try to hold on to some for a couple of months or longer and you'll be glad you did. With the amount of preservative honey and hops in this bragot, it will keep for many years if you let it.

Another option for a basic bragot recipe is to simply purchase grain or extract for a specific type of beer and add honey to make it a bragot. Some styles work better than others with honey. I've found that amber, pale, and wheat malts work best for bragots, as the bouquet and aroma of the honey really come out in them, but dark ales, stouts, and porters are also good candidates for flavoring with honey. The key is to practice and strike the right balance. Strive for complementary—not competing—flavors.

Brew in a Bag (BIAB) All-Grain Beer and Bragot

Once you've made a few batches of beer or bragot with an extract kit, you may begin to feel limited in your options for experimentation. While you can brew excellent beers from extract kits along with adjunct grains, hops, and herbs, brewing all-grain makes the avenues for experimentation almost limitless.

All-grain brewing can be incredibly complex, but it doesn't need to be. I brewed for many years without using all-grain equipment, and made some excellent beers from extract, some of them fairly experimental. I'm glad I finally made the switch, however, and encourage even beginning brewers to start out with all-grain. There are many books available on all-grain brewing, and tons of DIY all-grain systems discussed online. For the following all-grain recipes, I assume basic knowledge about the all-grain beer brewing process, as going into detail

Building an Immersion Wort Chiller

For many years I cooled my wort by buying a couple of large bags of ice, dumping them in the bathtub, and setting my brew pot in the ice bath, which often took far too long and made for late brewing nights. Once I learned how affordable it was to build a wort chiller, I realized that if I had made one in the beginning it would have paid for itself after just a few brewing sessions by saving the cost of that ice.

At its essence, a chiller is simply a way to quickly cool the wort by running cold water through copper tubing without lots of watching, worrying, and waiting. Building a wort chiller is so simple and afford-able, I can't believe it took me as long as it did. You can buy wort chillers for $60 to as much as $200 from homebrew-supply stores. Or you can build your own for as little as $25 to $30. While some of the pricier ones have fancy copper fittings and other pretties, all you really need to do is cool the wort, so why not save some money?

Be sure to shop around when looking for copper. While I initially planned on going with my local hardware store, I found the price per foot for their copper tubing was way out of reach. You may be able to find good prices at a plumbing store, but I found some Amazon Marketplace stores to be the best source. Thicker tubing will cool quicker, but any thickness of copper will cool quicker than an ice bath, so go with the best price you can find. I found the best pricing by purchasing precut 25-foot tubing with a wall thickness of 0.022 to 0.025 inch. Twenty-foot tubing will suffice, but try not to go with any less, and don't spend any more than $20 to $30. If you want faster chilling, you can use up to 50 feet of tubing, but I've found that my 20-foot chiller does the job much quicker than an ice bath so I don't see the need for the extra expense.

Materials
  20–25 feet (6–7.5 m) of copper tubing, ⅜- to ⅝-inch
      (9.5–16 mm) OD (outer diameter)
  8–10 feet (2.5–3 m) vinyl tubing, ⅜- to ⅝-inch (9.5–16) mm
      ID (inner diameter); be sure to match the inner diameter to
      the outer diameter of the copper tubing

3 hose-repair clamps, each ¼ to ½ inch (6–12 mm)
1 garden-hose female adapter (¾ inch [19 mm]) with a male
    adapter that will fit either the inner or the outer diameter
    of the vinyl tubing (I had difficulty finding any adapters
    with male ends under ⅝ inch (16 mm), so I found one
    that was ⅝ inch and thus allowed the tubing to fit
    snugly inside)
Tubing bender (optional)

PROCESS

1. Take one end of the copper tubing and pull it up through the center (it should come out of the box coiled). While I've heard of people mistakenly putting kinks in the tubing by doing this without a bender, I found that mine required little bending and wished I hadn't purchased the bender. You can do this around a cylindrical object of smaller diameter than your brew pot, but my tubing was flexible enough that I didn't find this necessary.

2. Place the hose-repair clamps around each end of the vinyl tubing, cut the tubing in half, and place one end of each half over the copper tubing. Then tighten the clamps.

3. Take the other end of the piece of tubing attached to the copper end-piece that you pulled up through the middle and work the garden-hose adapter around it (or into it).

4. When it comes time to chill your wort, simply take a garden hose, screw it into the adapter, and carefully set the chiller into your wort. The hot wort will sterilize the chiller. Before turning the water on, be sure you have checked for leaks at the connections. If you do have a bit of a leak, try to keep it from leaking into your wort. Be sure the end that the water will come out of is secure, as the force of water coming through it can cause the tubing to flail around and spray tap water into your wort.

5. Monitor the temperature with your thermometer or the back of your hand until you've reached yeast-pitching temperature, then shut the water off. I tend to use the excess water to rinse off cooking equipment, water my animals, or to fill up my washing machine.

is outside the scope of this book. Don't worry, though: Everything you need to get started is explained here. But don't let that keep you from reading up on all-grain brewing, either.

You don't need to purchase all the equipment you normally read is required for all-grain brewing. If you do any sort of cooking or grow and preserve your own food, you likely already have what you need in your kitchen. Just be sure to distract your domestic partner if you're going to borrow some cooking equipment—and reward their generosity with a tasty brew.

For this method, you'll be sacrificing some efficiency and may end up with fewer sugars than you would if you're going with the full setup. At least, that's what they say. Every beer I've made this way has been excellent and has been met with compliments from discerning beer drinkers. One trick is to add a pound or two of grain in addition to what the recipe calls for to provide additional sugars. For bragot, you're providing plenty of fermentable sugar in the form of honey, so I wouldn't worry too much.

EQUIPMENT
1 large 18- × 32-inch (46- × 81-cm) fine-mesh nylon
   straining bag

A turkey fryer with a propane tank is a worthwhile investment for brewing all-grain bragot and beer.

1 30- to 40-quart (30- to 40-L) stainless-steel or aluminum
pot (stainless steel is thicker and higher quality, but I
brewed for many years with aluminum)

1 15- to 20-quart (15- to 20-L) pot, a couple of additional pots
of varying sizes, and a tea kettle (optional but helpful)

1 colander or strainer large enough to set in the opening of
your brew pot

1 large stirring spoon (long enough to reach the bottom of
your brew pot)

1 high-temp brewing or candy thermometer

1 hydrometer (not absolutely necessary, but good to have if
you're shooting for a specific alcohol strength)

2 plastic fermenters (each 6 to 7 gallons [24 to 28 L]) with lids
and airlocks, *or* 2 glass carboys (each 6 to 7 gallons [24 to
28 L]) with corks and airlocks (or a combination of buckets
and carboys)

PROCESS

1. Clean and sanitize (see page 55 for a discussion on sanitization, and page 56 for sanitizer options) all equipment early

You can have grains for a bragot or honey beer ground at a homebrew
store, mill them yourself, or grind them in small amounts with a rolling pin.

on brew day or the night before, and allow equipment you won't be cooking with to fully air-dry.

2. If you're using a propane burner (gas heat is better for regulating temperature than electric), set it up outside or in a well-ventilated area. Be sure to get a solid blue flame going, as a yellow flame will cause a lot of soot on your pan—and a lot of grief from your housemate(s) or significant other.

3. Set the grain bag in the brewing kettle and fasten it around the edges with some hose clips or clothespins.

4. Fill the grain bag and loosely tie it, then soak the grains with water. For batches with 5 pounds of grain or less, you can add 1 to 1½ gallons (4 to 6 L) of water first. Beers such as stouts that can require as much as 10 pounds of grain will soak this water up quickly, risking burning the bottom of your pan, some of the grains, and the bag. It doesn't hurt to add a false bottom such as a canning rack to avoid this. Either way, add enough water so that the grains are covered with an inch or two. This is called the dough-in.

5. Heat the water to 160° F (71° C) and watch your thermometer until the temperature stays steady at around 145–155° F (63–68° C). Once it stabilizes, set a timer to 30 minutes and stir regularly, or lift the grain bag up and down periodically, moving the grains around with a spoon. Be gentle. You want the essence of the sugars from the grains to slowly work its way into the wort. If you press too hard, you may squeeze out too much tannin, which can lead to unfavorable flavors.

6. Start heating water in your kettle, and set aside a smaller pot. When the tea is boiling, pour it into the second pot and start another pot of tea water. Check the temperature to ensure they're both at about 185° F (85° C). You can use the second pot to refill your kettle, if needed.

7. When your timer goes off, it's sparging time! Set your colander over the opening of the pot and begin to slowly pour or scoop the mash (cooked grains) into it. Pour the water across as much of the surface of the grain bed as possible, moving the grains around with a spoon. This slow, measured pouring will ensure you sparge out the optimal amount of

sugars. You should end up with about 2 gallons (8 L) of water by the time you're finished, although more is fine.

8. From here, proceed with a one-hour boil, and add hops and honey as outlined in the specifications of the bragot recipes in this book or another recipe you're following. Note that some recipes may call for a 90-minute boil. Follow the hops schedule accordingly, but 60 minutes is the standard for the vast majority of recipes.

9. If you have a hydrometer (see page 91), check the gravity once the wort has cooled. For bragots, 5 to 8 pounds of honey should bring the OG (original gravity) to about 1.080. More honey can bring it to 1.100 or a bit higher. When the fermentation process is complete, check the gravity over a couple of days. It should stabilize at 1.010 or lower. If not, patience usually does the trick, although adding more yeast or nutrients should help finish up the fermentation. Subtract the final gravity from the original gravity, multiply by 105, and you'll have your final alcohol level. Many recipes call for a specific OG, but I'll admit I'm not always consistent about checking gravity for beer and bragot and rarely do with mead or wine. The results are almost always good, so why worry?

Herbal Bragots, Wild Honey Beers, and Other Avenues for Experimentation

As with any other mead or beer, the herbs, fruits, spices, and other flavorings you can add are limited only by your imagination (and knowledge of plant lore). After enjoying a couple of batches of a basic bragot, you'll likely want to begin experimenting. You can brew spiced bragots for holidays and other special occasions, or for medicinal purposes. Certain combinations work particularly well together, both for flavoring and for complementary health benefits.

While you can harness wild yeast to ferment beer in a similar manner to mead, a bit more care needs to be taken to ensure you lure in the friendlies. Barley and other beer grains may initially have yeast on them, but beer wort requires cooking at high temperatures and boiling for lengthy periods to draw out the appropriate amount of sugars and enzymes for producing alcohol. Thus, you are killing off any wild creatures that may be residing on the grain. I still often use commercial yeast when brewing beer to a particular style, but venture into wild-fermentation territory when I'm not concerned about style and don't mind a bit of a sour or tart edge. This is yet another reason why I often use honey as a flavoring and bottle-priming agent for beer, as adding unpasteurized honey will introduce additional wild yeast and bacteria. You'll still need to do a bit more to ensure a healthy yeast population, though.

Sometimes when I brew beer, I set aside a gallon or two (4 to 8 L) of wort in an open fermenter before pitching the yeast in my main fermenter. I'll also sparge a second or third time after gathering the initial runoff from my main batch and make lighter beers from the additional wort. This was a common practice in traditional farmhouse brewing, and with commercial brewing up to the early 1900s.[2] One term for these additional batches (specifically, every batch following the second) is *small beer*. There would be fewer sugars in this beer, meaning less

When making bragot, ale, or beer, you can scoop grains from the first sparging and run more hot water through them to gather enough wort for additional "small" beers. This is also known as the parti-gyle method.

alcohol and thus a narrower window for drinking it before it spoiled. Small beers provide an opportunity to have a refreshing beverage or three after a day of working in the garden or fields. Another term for small beer is *parti-gyle*, as described by Randy Mosher in *Radical Brewing*: "Several hundred years ago, beers were mashed in two, or more often three, mashings, with successive infusions resulting in weaker and weaker beers. Each beer was run off to its own fermenting vat, or gyle."[3]

If you want to introduce wild yeasts to your small beer, or to your main batch, some options are to add raw honey (a cup or two will do if you're not making a bragot), 1 to 2 cups of organic raisins, or plenty of fresh fruit. Belgian brewers to this day employ wild yeast in brewing their beers, albeit on a fairly sophisticated level based on generations of brewing using the yeasts of a particular region. This isn't to say you can't make your own palatable wild brews if you're willing to experiment and set aside the modern cultural (at least in North America) obsession with sanitization and beer styles.

Strangely, although books and recipes on crafting wild beer exist, they generally recommend adding "wild" yeast strains that have been gathered from Belgian breweries and packaged for sale. These strains can still produce wildly divergent and complex flavors, and lessen the likelihood for "error," but this isn't truly wild brewing in my mind. That being said, I highly recommend Jeff Sparrow's *Wild Brews: Beer Beyond the Influence of Brewer's Yeast*. The author goes into great detail on wild beer history and technique, and provides plenty of information on how to craft your own wild brews in the Belgian brewing tradition, with or without store-bought yeast.

You don't need to follow complex formulas and pitch specific wild yeasts to experiment with wildcrafted beers. Ancient cultures used various techniques to ferment their beers (as do indigenous cultures today), and very likely used honey to initiate fermentation before they discovered other avenues for coaxing sugars from the starches present in grains. Saliva contains an enzyme that can convert starches to sugars, which is why chewing grains was also often part of the process.[4] In the Finnish epic the *Kalevala*, we learn of various attempts to procure yeast for fermentation in a story about the origins of Finnish beer. These techniques ranged from evergreen fir (along with evergreen bark, still an integral ingredient in traditional Scandinavian brewing) to the saliva of "enraged bears" (thankfully, no longer a critical ingredient). The last ingredient added—honey—was what finally initiated fermentation.[5] Rather than trying to gather the foam dripping from the jaws of a pissed-off bear, or convincing your friends to sit for hours to chew grain and spit it into a cauldron, you have some other options available.

Honey is clearly one of the core elements for initiating a wild fermentation—just add raw honey or honey from a batch of fermenting mead to your beer wort once it is "blood warm." Fresh, pesticide-free berries and fruits will also help get things moving along. Cherries have been used in wild Belgian ales for centuries, and other fruits have also become prevalent over time. Although you can use pretty much any fruit, those more common in Belgian wild beers include raspberries, wine grapes, apricots, black currants, peaches, and strawberries.[6] Each fruit has its own amount of sugars, tannins, and acids, but don't fret too much over this. Just use what's available seasonally. Pick fruit fresh and freeze it until you're ready to brew. Store-bought fruit, even organic, is often picked before its prime, and will come from questionable sources. While it will work as a flavoring agent, try to avoid fruit (fresh or frozen) from grocery stores if you want to get the optimal effect.

As with wildcrafted mead, be sure to start out with small batches until you've worked out techniques and flavor combinations that work best for you, and keep an open mind when tasting. Some wild beers taste great when they're first brewed, and others require aging (from a few months to several years) for their flavors to really shine. Keep in mind that Belgian beers are specific to the region where they are brewed. Although you can attempt to emulate them, the goal here is to create a brew unique to your own region and its wild yeasts.

As with mead, a common tradition in brewing beer—which still holds true in Belgium today—is to blend brews of various flavors and ages to enhance the flavor or mitigate unfavorable characteristics. If you want to emulate this process on a smaller (and quicker) scale, set aside a portion of a beer you're brewing and leave it out in an open fermenter with a cloth covering it for about a week. Taste it; if it's soured, mix it back into the entire batch, or add a bit to each bottle if you're not feeling adventurous. This is actually rumored to be one of Guinness's secret "ingredients" (or rather, techniques). I sometimes like to add 1 pound of acidulated malt (also known as sauermalz or sour malt) to my grain

Asheville, North Carolina's Wicked Weed Brewing ages its "funky" wild ales in used bourbon and wine barrels at its Funkatorium facility. This process imparts wild flavors that have soaked into the oak from previous ferments. The Funkatorium offers several sour ales and blended Belgian-Lambic-style ales on tap.

mix when doing a second sparge for a small beer. This malted barley contains a small proportion (about 1 to 2 percent by weight) of lactic acid, which can be used to lower the pH levels of high-alkaline water, or to intentionally add a sour edge to beer.

Big Yeti and Little Yeti: Barley-Wine-Style Bragot and Welsh-Ale-Style Small Bragot

As discussed on page 158, a good way to take full advantage of an all-grain or partial-grain brewing session is to create one strong beer, ale, or bragot, and one or more "small" batches by sparging multiple times. I find this works well when I want to make a barley wine, which is a strong pale ale (about 11 to 12 percent ABV) made from concentrated wort. Barley wine is a great beer to make into a bragot, as it is a heavy, malty beer already, and honey tends to balance well with the malt. It is generally very hoppy, but was traditionally made with more herbs than hops (if any). Whether you add hops or not, barley wine needs a year or more of aging for the strong flavors to mellow out. This is another reason to make a lighter beer or bragot along with it, so that you can have several additional bottles to hold you off while it ages.

Homebrew guru Randy Mosher's books *Radical Brewing* and *The Brewer's Companion* are great resources if you want to get into the technical details; however, this is a fairly simple process if you do seat-of-your-pants brewing as I do. My recipe and technique for Big Yeti was modified from Mosher's Big Stinky & Little Stinky: A Basic Parti-Gyle Recipe.[7] Little Yeti is a variation on some Welsh-style honey ale and gruit recipes I've come across.

GRAINS

4 pounds (2 kg) Muntons pale ale malt
1 pound (.5 kg) Gambrinus honey malt
1 pound (.5 kg) Briess caramel malt
4 pounds (2 kg) Munich malt
1 pound (.5 kg) dark crystal malt

HOPS FOR BIG YETI

2 ounces (57 g) Cascade (60 minutes)
2 ounces (57 g) US Fuggle (60 minutes)
1½–2 ounces (42–57 g) Cascade (30 minutes)
1 ounce (28 g) Sorachi Ace hops (end of boil or dry hop)

HOPS FOR LITTLE YETI

1 ounce (28 g) Cascade (60 minutes)
½ ounce (14 g) US Fuggle (30 minutes)
½ ounce (14 g) US Fuggle (end of boil)

HERBS & SPICES FOR LITTLE YETI

1 ounce (28 g) mugwort
2 teaspoons crushed coriander
1 teaspoon (5 mL) caraway
1 teaspoon (5 mL) allspice
1–2 teaspoons (5–10 mL) ginger, grated
½ teaspoon (2.5 mL) fenugreek
½ teaspoon (2.5 mL) black pepper
Zest of 1 orange

HONEY & YEAST

3 pounds (1.5 kg) medium-amber honey (for 3 gallons [12 L] of Big Yeti)

2 pounds (1 kg) medium-amber honey (for 5 gallons [20 L] of Little Yeti)

1 packet (5g) English ale yeast or 1 cup (249 mL) barm (for 3 gallons [12 L] of Big Yeti)

1 packet (5g) English ale yeast or 1 cup (249 mL) barm (for 5 gallons [12 L] of Little Yeti)

Use the basic all-grain process outlined in this book or a standard all-grain process if you have the appropriate equipment. Either way, take about 3 gallons (12 L) from the first sparge and put in in a stock-pot. Sparge the grains a second time and gather enough liquid to brew 5 gallons (20 L) into your larger stockpot (you can add water to the fermenter to make up for any excess if you can't brew 5 gallons [20 L]).

A blow-off tube is a good idea for the early stages of fermentation when brewing bragots, melomels, or any other brew with a vigorous fermentation.

Proceed with each boil separately. Be sure to keep the ingredients and process separate, although I do sometimes add a bit of the wort from the Little Yeti to the Big Yeti to reach 3 gallons (12 L). I once had a marathon brewing weekend that resulted in three 5-gallon batches of mead, a 3-gallon Big Yeti, and a 5-gallon Little Yeti. If you plan right and make sure you have all your ingredients and equipment on hand, this is a great way to really get a head start on stocking your cellar with tasty mead and beer.

Booby Bragot: A Lactilicious Bragot

My wife, Jenna, and I have used the same pediatrician for both of our daughters—one of the reasons being that she tends to prefer natural treatments over pharmaceuticals or antibiotics except when absolutely necessary. Another reason we like her is because she recommended Jenna drink one stout beer at about four o'clock every afternoon to aid with milk production. Hops and grains are both galactagogues (lactation enhancers), and stouts generally have plenty of both. The reason for drinking a beer at four was because that's generally when the human body is at its lowest level of activity physiologically, so the galactagogues can get to work without too many interruptions. Of course, the wheels started turning when I heard this. I decided to look into some herbal galactagogues, as I was preparing to make an all-grain rye pale bragot and figured I may as well get some brownie points while I was at it. It turned out to be crisp, hoppy, and very flavorful. For the next batch, I pulled out all the stops and made a thick, robust oatmeal-honey stout.

I can't say how much it helped get the milk flowing, but Jenna definitely produced more with our second daughter than with our first. This works just as well as a manly beer—I was worried it might cause me to lactate, but it didn't. When I first made it, I did it as a 5-gallon all-grain and fermented a couple of gallons separately as a small beer. I made a tea from the herbs and berries and added the tea to the small beer. I also added some to the main batch, but the flavor profile from the herbs was hardly noticeable. If you're more interested in the effect than the flavor, add the herbal tea at the beginning of the boil or halfway. If you taste the tea and like its flavor, add it in the last five minutes or after the boil is complete.

This bragot, as with any other, can also be fully or partially wild-fermented to give it a bit of a tart or sour edge. When I want to do this, I open-ferment about a gallon (4 L) of wort from a 5-gallon (20-L) batch and leave it for a couple of weeks to sour a bit. I also add some acidulated malt (see page 160) to the grain bill. To be honest, it tries my taste buds, but my wife loves it. She's one of those people, though, who can bite into a lemon or lime and actually enjoy it. My friend and Viking-mead coconspirator Steven Cole once tried some and had a reaction similar to mine: "Tastes like sour milk mixed with crab apples." If this sounds good to you, open-ferment to your heart's desire and use acidulated malt. If not, leave the acidulated malt out and don't open-ferment. I also sometimes leave hops out as Jenna isn't a big fan of bitter flavors. Every other aspect of it is delightful, though. It has a thick, almost syrupy body, and the other flavors blend nicely with the tart—if that's your thing. If you're more into bitter flavors, add some bittering agents and you'll be pleased. When I brew it with hops and without sourness, it makes for an outstanding beer that would rival any modern commercial stout. Keep in mind, though, that nearly all ancient beers had a tart or sour edge to them, so a bit of sourness is actually very historically accurate.

GRAINS FOR LITTLE BOOBY BRAGOT

8 pounds (4 kg) pale ale malt (for extract and all-grain brewing); *or* 6 pounds (3 kg) amber dry malt extract (for extract brewing)

2 pounds (1 kg) Vienna malt (for extract and all-grain brewing); *or* 2 pounds (1 kg) pale dry malt extract (for extract brewing)

GRAINS FOR BIG BOOBY BRAGOT

8 pounds (4 kg) Munich or pale ale malt; *or* 4 pounds (2 kg) pale dry malt extract (for extract brewing)

2 pounds (1 kg) amber malt; *or* 2 pounds (1 kg) amber dry malt extract (for extract brewing)

1 pound (.5 kg) black patent malt (for extract and all-grain brewing)

1 pound (.5 kg) dark crystal malt (for extract and all-grain brewing)

½ pound (225 g) chocolate malt (for extract and all-grain brewing)

½ pound (225 g) roasted barley (for extract and all-grain brewing)

1 pound (.5 kg) acidulated malt (for extract and all-grain brewing; optional)

ADDITIONAL INGREDIENTS

6–12 pounds (3-6 kg) light to medium honey such as sourwood, wildflower, tulip poplar, or tupelo

1 cup (240 mL) blackstrap molasses

1–2 cups (240–475 mL) oatmeal (preferably rolled or
old-fashioned, but instant will work)

2 packets ale yeast or 1 cup barm

5 tea bags of organic Traditional Medicinals Mother's Milk tea;
or as many of the following herbs as you are able to locate:

> 1 ounce (28 g) fennel seed
> ¼ ounce (7 g) anise
> 1 ounce (28 g) coriander
> 1 ounce (28 g) fenugreek seed
> 1 ounce (28 g) blessed thistle
> Pinch of spearmint leaf
> Pinch (a careful one) of nettle leaves
> Pinch of West Indian lemongrass leaf
> Pinch of lemon verbena leaf
> 1 ounce (28 g) marshmallow root
> 2–3 ounces (57–85 g) fresh grated ginger or
> ginger powder

OPTIONAL BITTERING
AND PRESERVATIVE AGENTS

1–2 ounces (28–57 g) Cascade or any other high-alpha hop

½ teaspoon (2.5 mL) mugwort (use with caution; some sources
say it helps lactation, and others say nursing mothers should
avoid it due to its neurotoxic properties)

PROCESS

1. Pour ½ gallon (2 L) water into a small stockpot. Add the tea
   bags to the water along with any additional herbs or spices
   in a tea ball or cheesecloth—or simply add them loose and
   plan on filtering later. Bring the tea to a light boil, simmer
   for about 30 minutes, and cut off the heat. Do this at any
   point during the brewing process, or even the night before.
   Be sure to give it enough time to cool to blood temperature.

2. If you're brewing with extract, proceed as with the extract recipe on page 147, adding additional crushed dark grains and oats in a grain bag to steep.

3. For an all-grain batch, pour all the grains into a grain bag in your brewing kettle, add enough water to cover the bag, and bring the water to 160° F (71° C). Stabilize at 150° F (65° C) and keep it at 145–155° F (63–68° C) for 30 minutes.

4. Bring the water up to 175° F (79° C) and sparge with about 2½ gallons (10 L) of 185° F (85° C) water.

5. Transfer the wort to a brewing pot, add the molasses, and boil for one hour.

6. Add the hops and/or mugwort halfway through the boil.

7. Cut off the heat, add the honey and tea, and stir well.

8. Cool to "blood warm," transfer the wort to a fermentation vessel with an airlock, add yeast, or go wild.

9. Wait impatiently, rack in about a week, and continue to twiddle your thumbs.

10. In about another week, you should see no further sign of fermentation. Prime the wort with ½ cup (120 mL) honey dissolved in ½ gallon (2 L) warmed water, bottle, and try very hard to wait at least two weeks to allow it to fully carbonate. Drink at any point for lactation assistance.

11. Have a booby-beer party (not *that* kind of booby-beer party), or savor quietly with your boobies (or those of your significant other).

Herbal Sorghum Beer

When I was attending to some yard work one hot summer day, I realized that my yard was more of a garden than a traditional yard. Considering the plants my wife and I grow in our garden plots and raised beds, our fruit trees and berry bushes, our mushroom logs, and the wild edible "weeds" that pop up throughout the spring and summer, most of our yard is edible. After contemplating this, I realized that what I was doing was "yardening." (I'm not the first to come up with this word—I discovered later that other gardeners who live in urban areas have used it as

Straining some sorghum yard beer made with
purple dead nettle and creeping charlie (also known
as alehoof).

well.) When I make beer and mead using ingredients from my yard
and garden, I like to call them yarden beers or yarden meads. Here is a
recipe for a yarden sorghum beer or, if you prefer, a yard beer.

You can use this as a blueprint for making all manner of herbal
beers. See chapter 2 for a list of traditional brewing herbs and their
medicinal benefits. Because it's made from sorghum (or molasses as a
substitute), it's also gluten-free. For many years, sorghum and molasses
were common bases for beer recipes when grains weren't available, and
all manner of herbs were used to flavor and preserve them. So keep in
mind when you're drinking this that you're imbibing a bit of history.

INGREDIENTS FOR 1 GALLON (4 LITERS)

1 pound (.5 kg) sorghum, molasses, or malt extract from a
homebrew kit

1–2 cups honey (240–475 mL)

1 gallon (4 L) water

2 teaspoons (10 mL) ale yeast, ½ cup (120 mL) barm or ginger
bug, or wild ferment

3–4 sprigs alehoof/creeping charlie

1–2 sprigs purple dead nettle

½ cup (120 mL) (give or take) wild violet petals

½ cup (120 mL) (give or take) dandelion petals

10–12 raisins

1 cup (240 mL) oatmeal (optional, for adding body)

PROCESS

1. Pour 1 gallon (4 L) of quality dechlorinated water into
   a stockpot. Heat to "blood warm" or about 70–80° F
   (15–27° C). Pour 1 pound (.5 kg) of sorghum extract into
   the water, stirring constantly. Shut off the heat and stir in
   the honey.

2. Stir in the oatmeal, raisins, and desired botanicals. You can
   substitute other herbs for those listed above, or add more to
   make a powerful herbal tonic.

3. Add yeast, barm/bug, or wild-ferment.

4. I usually open-ferment this beer, even when adding yeast
   or barm, so that I can stir it a couple of times a day to
   aerate and allow the essence of the herbs to work its way
   into the brew.

5. After about a week, strain out the herbs while transferring
   the batch to a 1-gallon (4-L) jug with an airlock. You will
   now have imparted into the beer the medicinal qualities and
   wild yeast from the botanicals. If you want to enhance the
   flavor with a plant that appeals to you, add it now.

6. Age for a week or more to ensure it is fully fermented,
   bottle, and carbonate if desired.

7. Start drinking in a week or two or give it as much time as
   you desire to age.

Horehound Beer: Ye Olde Variations

The 1888 book *A Treatise on Beverages; or, The Complete Practical Bottler* by Charles Herman Sulz has a couple of variants on horehound beer worth experimenting with.

> ### Horehound Beer
>
> To make ten gallons, make an infusion of two ounces of quassia with two dozen sprigs of horehound; boil with part of this liquid thirty cayenne pods for twenty minutes, then add ten fluid ounces of lime juice and two ounces liquorice (dissolved in cold water); strain the mixture and put with it ten gallons of cold water, with three pounds brown sugar, caramel to color; allow the whole to work four days. Now take four quarts of it, and warm it to the proper temperature, and mix with this one pint of good brewers' yeast, and stand it in a warm place till in a brisk state of fermentation; mix it with the rest of the liquor, and in a few hours it will be all in full fermentation. Give it a stir twice a day for the first two days to promote fermentation; keep it from contact with cold air for the

Horehound Ale

This is a traditional tonic ale that was popular throughout Europe for many centuries, and is indigenous to Britain. In her manual *A Modern Herbal*, Mrs. Maud Grieve references it as "an appetizing and healthful beverage, much drunk in Norfolk and other country districts."[8] Here's a recipe I came up with based on several variations I've read.

INGREDIENTS FOR 1 GALLON (4 LITERS)
1½ ounces (42 g) (Some recipes call for as much as 30 ounces per gallon!) dried horehound or 2–3 sprigs fresh
½ ounce (7 g) crushed coriander seeds

following two days, and skim the top off as it gets yeasty. In thirty hours the beer may be bottled off. In summer this will be ripe and fit to drink in eight days. A superior quality may be made by putting a small piece of sugar into each bottle just before corking.

Another formula: water, ten gallons; sugar, five pounds; horehound herb, ten ounces; camomile, two ounces; Jamaica ginger, bruised or crushed, six ounces; good fresh yeast, one pint; liquorice [extract] for coloring, one ounce. The latter made into a liquor with a pint of boiling water. Put the horehound, camomile and ginger in an open gauze or coarse flannel bag, and let them together boil gently for two hours or longer, to extract all the aroma from the herbs and ginger; then remove all the liquor into a tub or large pan, and at about eighty degrees of heat add the yeast. Stir the mixture, and let it stand with a cover over it for ten or twelve hours, after which put it into a cask to ferment, taking off the yeast as it arises at the bung-hole. This preparation is made stronger by adding an ounce of the extract of malt mixed with the liquor when cooling.[9]

½ ounce (7 g) dried ginger or 1 ounce (28 g) grated
2–3 cups (475–700 mL) honey
2 pounds (1 kg) dark malt beer extract, molasses, or sorghum
1 gallon (4 L) water
1 teaspoon (5 mL) brewer's yeast or ale yeast, or ½ cup
    (120 mL) barm
10–12 raisins

PROCESS

1. Make a tea by steeping the horehound, coriander, and ginger in 2 cups (475 mL) of hot (boiled, then removed from the heat to cool) water for one hour.

2. Strain the herbs, then bring the water back to approximately 60–70° F (15–21° C), or "blood warm."

3. Pour in the rest of the water and stir in the honey and malt extract, sorghum, or molasses to dissolve.

4. Add the yeast or barm and raisins, ferment, prime bottles, fill, cap, and drink in 1–2 weeks.

Blending Your Brews into an Ancient Grog

As discussed in chapter 2, the Norse and other ancient cultures often blended various fermented beverages together to create a grog. When you get to the point of having several meads, beers, and wines brewing at once, this is something you can experiment with. When I rack and bottle my beer, mead, and wine, I usually spend most of a day or even a good bit of a weekend working on first, second, and third rackings, and bottling anything that's ready. Sometimes I'll end up with a gallon or two extra if I'm switching between different-sized carboys or buckets. Other times I'll intentionally take out a gallon and replace it with a batch of fresh honey-water (or fruit juice if I'm making a melomel). When I do this, I take a bit of each, mix it in a beer stein or growler (½-gallon jug used to transport draught beer), and sip on it while I'm working. When I create a blend I particularly like, I take any excess I have, blend it in a carboy or fermentation bucket, and add enough honey water to fill it to the top. I've even gone as far as to mix several of my meads, a bit of bragot, and some wine and mead from a commercial winery. This will turn out different every time you do it, but nearly always makes for an enjoyable drinking experience. It's not beer; it's not wine; it's not mead—it's grog!

Sima: Finnish May Day "Mead"

A traditional Finnish beverage made to celebrate the arrival of warm weather, sima is more a sparkling lemonade than a mead. Although it was made with honey in ancient times, nearly all recipes today call for water, sugar, lemons, raisins, and yeast (brewer's or bread). You can also spontaneously ferment this or use ginger bug.

INGREDIENTS FOR 1 GALLON (4 LITERS)

1 gallon (4 L) water (give or take)

2 large lemons

½ cup (120 mL) brown sugar

½ cup (120 mL) cane sugar or 1 cup (240 mL) honey

¼ teaspoon (2 mL) dried yeast, ½ cup (120 mL) ginger bug,
    or wild-ferment

25–40 raisins

PROCESS

1. Bring the water to a boil.
2. Peel off outer yellow lemon rind (zest) and discard the inner white rind.
3. Place the zest, sugar, honey, and lemons in a 1- to 3-gallon (4- to 12-L) fermentation vessel.
4. Once the water boils, turn off the heat. Pour directly into fermentation vessel if you're using a heatproof vessel. Otherwise, allow it to first cool to room temperature.
5. Once the must is at room temperature, add the yeast or barm. If wild fermenting, add 25 raisins (separate from the raisins you will add later), and stir several times a day until fermentation has commenced.
6. Once you see light bubbling, strain and rack the liquid into thick-glass bottles, preferably swing-top. Add 2 to 3 raisins and ½ teaspoon (2.5 mL) of white sugar to each bottle.
7. Allow the bottles to sit at room or cellar temperature for 12 to 24 hours, then *carefully* place them in a refrigerator. In two to five days, they should be ready to drink, which should be indicated by most of the raisins having floated to the top of the bottle.
8. Open carefully and enjoy. If the flavor is too yeasty or lemony, mix in some fresh honey or simple syrup (sugar dissolved in hot water and then cooled). Don't forget about the bottles! Remember, this is heavily carbonated and will continue to ferment while in the bottle.

CHAPTER NINE

How to Drink
Mead Like a Viking:
Viking-Era Games and Rituals

The man who stands at a strange threshold,
Should be cautious before he cross it,
Glance this way and that:
Who knows beforehand what foes may sit
Awaiting him in the hall?

—Hávamál *(Sayings of Har/Sayings of the High One)*[1]

Wile a lot of pleasure and work (and a fair share of frustration) go into making mead, the obvious goal is to end up with a product that you will enjoy drinking. The experience of drinking some meads is akin to that of wine, while others are more like ale or beer. You can certainly enjoy sipping mead slowly out of a wine-glass while having a romantic dinner in front of the fireplace with your mead wench (or fella), but there are many other resources on mead, wine, and beer that discuss how to drink in this manner. Rather, in this chapter, we'll delve into the significance of drinking rituals in Norse and Anglo-Saxon cultures, and the recreational activities Vikings took part in while enjoying their mead. First, though, I suggest stepping back and thinking about exactly how the production and consumption of alcohol should be pursued.

Thoughts on Enjoying Alcohol in Moderation

Less good than belief would have it
Is mead for the sons of men:
A man knows less the more he drinks,
Becomes a befuddled fool

—Hávamál *(Sayings of Hár)*[2]

To understand the modern fascination with alcoholic beverages of all kinds, as well as the reasons why they are also targets of condemnation, we need to step back and take a longer view. Alcohol occurs in nature, from the depths of space to the primordial "soup" that may have generated the first life on Earth. Of all known naturally addictive substances, only alcohol is consumed by all fruit-eating animals.

—Patrick McGovern, Uncorking the Past[3]

Our ancestors had a connection with the tiny creatures responsible for bringing food and drink to life in times of plenty—and preserving both to keep them going through the difficult times when nourishment was scarce—that was an integral part of their daily lives. Even in the early days of the industrial revolution, people were largely reliant on what they could produce or barter for with their friends and neighbors.

For the most part, fermentation in early societies was simply a way to create and preserve food to enable a community to not only survive, but also thrive. Often a natural by-product of this process—for beverages—was alcohol. Unless allowed to ferment for long periods of time, most alcoholic beverages created for daily consumption had such low levels of alcohol that they allowed people to work in the fields (see the discussion on small beers and parti-gyle brewing in chapter 8), or accomplish whatever other contributions to society they took part in without falling over in a drunken stupor halfway through the workday. High-alcohol beverages such as wine, ale, and mead were brewed and aged for special occasions and were generally consumed during community or family gatherings, either in small amounts with the evening meal, or in large amounts to celebrate events such as weddings or the end of a successful harvesting season. This isn't to say that over-indulgence didn't happen, nor that high-alcohol beverages weren't produced and enjoyed—we have evidence of extreme drunkenness in the Norse sagas, and in other ancient sources. Additionally, as discussed in chapter 2, highly intoxicating fermented beverages were created for both shamanic and recreational use with plants that had powerful effects of their own that surpassed even those of alcohol. But alcoholism, as we know it today, is a fairly modern construct that came about as consumerism led us away from producing our own alcohol and made it easy to simply go to the store and buy as much booze as we want. Over time, as long-term aging and distilling were perfected, beverages of higher and higher alcohol content became simple for practically anyone to make, resulting in increased instances of over-indulgence and the attendant societal implications.

In turn, religious and government figures began to notice both the enjoyment and problems that resulted from this over-indulgence and—with understandable but misguided intentions—began to create laws that attempted to regiment the production and intake of alcoholic and botanical intoxicants. With prohibition comes further over-indulgence, just as with over-abundance comes abuse. As alcohol and other intoxicants became more readily available to the common person, ritual and long-standing family traditions fell by the wayside and misuse of alcohol became more prevalent. However, as America learned through the dark days of Prohibition and the War on Drugs, prohibiting outright a substance that causes any degree of pleasure simply gives enterprising individuals a new route (legal or otherwise) for exploiting those who wish to partake.

As I write this, the corner of eastern Kentucky where I reside is slowly rolling back the fetters of Prohibition. While there are several wineries, distilleries, and breweries within a fairly short drive of my house, and I can drive 15 miles to purchase alcohol, I live in a dry county. What this means is that, while I can legally purchase alcohol elsewhere and bring it home with me, and can legally produce small amounts of nondistilled alcohol for personal consumption, no restaurant or other business is permitted to sell alcohol in an area that has been legally mandated as "dry." Dry communities surrounding me over the past couple of years have begun to pass laws turning their counties "moist" (alcohol can be sold in restaurants, but not stores) or even "wet" (booze for everyone!). This has not been an easy task. Many people, even those who indulge in alcohol, are up in arms about it. The plain and simple fact of the matter, though, is that people will always consume alcohol. While I would prefer that more people learn how simple it is to make their own, there is something to be said about being able to go out for dinner and have a glass or three of wine, beer, or even mead with friends and family. What it all comes down to is personal accountability and the freedom to decide for oneself whether or not to purchase alcohol, and to consume it in moderation. By no means do I intend to make light of the very real problems of addiction, and the myriad other problems that alcohol can cause. However, what I hope to do is present an alternative to the all-too-common tendency to drink for the sole purpose of getting soused. As someone who has done this many times in the past, I would be hypocritical to imply that I don't perfectly understand the temptation and joys (at least the night before . . .) of doing this. However, as I've begun brewing more of my own alcoholic beverages, and taking the time to sample well-crafted beer, mead, wine, and liquor, I've come to realize that "chasing the buzz" shouldn't be what it's all about. When taking the time to slowly drink and truly enjoy a beverage that has been fashioned with care—particularly in the company of others who are also willing to take the time to savor the fruits of the bottle—it's possible to enjoy a buzz on an entirely different level. While the alcohol may have some influence on this buzz, the combination of a naturally induced fermentation from botanical ingredients and unprocessed sugars (in other words, honey) and all the benefits this imparts, along with good company and conversation . . . well, let's just say that when you reach a state where all the truly good things in life come together, it's worth savoring the moment. Whether we make our own alcohol or purchase it at a store or restaurant, well-crafted alcoholic beverages can greatly enhance community and family events. This is how alcohol was enjoyed in the past and how it should be continued to be enjoyed into the future.

Mead and Ceremony

Due to the sacred and mystical nature of the process of fermentation to the Norse and Anglo-Saxons, the drinking of mead wasn't always taken lightly. While it did play a large part in recreation and was certainly

A reconstruction of the world's largest Viking longhouse at Lofotr Vikingmuseum in Borg near Bostad, Norway. It was likely the seat of a powerful Viking chief. At a length of 272 feet, it contains authentic re-creations of both a mead hall and living quarters. Photo courtesy of Paul Berzinn, Wikimedia Commons.

enjoyed before and after battle and other significant events, the actual act of initiating a drinking session was often shrouded in ceremony. Norse culture was very hierarchical; each person had a place in society, and matters such as what they drank, when they drank, and how they drank were integral to keeping order. If the set rituals weren't followed properly, fights to the death or severe punishment could result. On the flip side, as long as everyone followed the rules, or invoked a challenge without straying too far from custom, Viking and Anglo-Saxon feasts could be raucous, epic occasions.

There were two primary types of feasts that would take place in a mead hall: the *symbel*, a highly formal type of feasting, and the less formal *gebeorscipe*. For both, "strong drink" was a core component.[4] For either type of feasting, the layout of the mead hall helped to make each person's place clear. The mead benches, situated at the lower end of the hall, were where the freemen—the chieftain's willing followers—engaged in mostly unfettered drinking and camaraderie. They recalled glories and defeats on the battlefield, told stories, and boasted. The fellowship in this part of the hall served another important purpose—that of steeling warriors for battle. When traveling by sea or land for raiding, diplomacy, or trade—or while on the battlefield awaiting the order to attack—memories of feasting in the warmth and security of the mead hall bolstered them for what was to come. These could be lengthy, brutal affairs and often involved many months, or even years, away from the hearth of home and the comfortable accommodations of the mead hall. On a raised dais at the upper end of the hall sat the chieftain (or lord, king, jarl/earl, what have you), his lady (or queen), and their royal retinue. Speech was "relatively formal and carried official weight—feasting here was as much about discussing serious business as about entertaining the fellowship."[5] This area would also serve as a court where persons of any status could bring matters to be discussed, after which the lord would deliver the verdict.[6]

In *Lady with a Mead Cup: Ritual, Prophecy and Lordship in the European Warband from La Tène to the Viking Age*, Michael J. Enright goes into great detail explicating the rituals of drinking mead at feasts—and the important role of women in drinking rituals—drawing primarily from references in *Beowulf* to the *symbel* that takes place in *Heorot*, the great mead hall of King Hrothgar. Though it is an Anglo-Saxon poem, *Beowulf* has strong historical and literary connections with the Norse, and provides a first-rate example of what life was like for both the Vikings and Anglo-Saxons (with the caveat that it is also an early example of literary hyperbole). Upon arriving in Heorot, Beowulf is "given an honorable seat between the king's sons and witnesses the entrance of Wealhtheow, whose formal offering of the ceremonial cup of liquor to the Danish king signals the true beginning of the feast."[7] As we see in the poem itself:

> . . . Stepped forth Wealhtheow,
> Hrothgar's queen, careful of noble usage.
> Gold-adorned she greeted the men in hall.
> And then the gracious woman offered the cup
> first to the East-Danes' king, the guardian of their land,
> bade him be happy in partaking of this beer-assembly,*
> be amiable to the people. He took part eagerly
> in the feast and formal cup, fully victorious king.
> Then her steps led here and there the lady of the Helmings,

* The words for "mead" and "beer" were often used interchangeably.

To veterans and youths, to each group of them,
she offered from the treasure-vessel. Until the moment came
when to Beowulf the ring-adorned queen,
gratified in her heart, brought the cup of mead.
She greeted the chief of the Geats, gave thanks to God,
wise as she was in her words, that her wish had come to good
that she might have confidence in some hero,
a comforter in her woes. He partook of the cup,
the fierce slayer, from Wealhtheow's hands.
And then chanted his eagerness to fight.[8]

There is an obvious order to the ceremony. First, the queen greets the warriors, but offers them no mead. Instead she presents the cup to the king and bids him enjoy the drinking and be happy with his people. And enjoy it he does. If Wealhtheow were to next haphazardly offer the cup to a younger warrior, the elder warriors (and by extension, the king) would be greatly offended, and bloodshed would likely ensue. Instead, she offers the cup first to the veterans and then to the younger and unproven warriors. Beowulf—being an honored guest—is presented with not only a drink from the cup, but also a speech honoring him—to which he responds with a speech of his own.

Although not all feasts were this formal, every element of a *symbel* was fraught with long-standing rituals, and each held its own weight—not least of which was the drinking vessel and the drink within. Ceremonial mead was not just any mead, but one brewed special for the occasion from the finest of ingredients. Even events that started out with high levels of pomp and circumstance often turned quickly to feasting, merrymaking, boasting, storytelling, verse giving, game playing, and much mead swilling once the formalities were complete. Still, each person participating in the feast recognized their place, or was at least expected to. Sometimes, guests would intentionally dishonor their hosts or another member of the party if they felt they had been wronged in some manner or simply wanted to stir up trouble.

The *gebeorscipe*, which can be translated as "beer drinking" or "drinking party," was where the real fun took place. Given the ambiguous nature of the words for "mead" and "beer," let's just go ahead and call it a mead fest. Mead fests were less formal events than were *symbels*. Likely,

A scene on a stone from Gotland, Sweden, on display at the Swedish History Museum in Stockholm. The image represents participants in drinking rituals that were common in formal feasts called *symbels* and less formal feasts called *gebeorscipes*.
Photo courtesy of Berig, Wikimedia Commons.

these were smaller-scale events held in individual houses, or larger (but still informal) feasts held in the communal mead hall.

Although impromptu drinking feasts surely occurred—particularly in the long winter months—an official feast was expected to last a full day. Special events were sometimes held to showcase the wealth of the host or to mark a momentous occasion, and were required to last three full days. If food or drank ran low before the end of the third day, this showed very poorly on the host.[9] Many examples of feasts similar to the descriptions of *gebeorscipes* and *symbels* are referenced in the Icelandic sagas, indicating that communal feasting and drinking events were not uncommon.

Games for Gods and Midgardians

In addition to drinking festivities, game playing was a common way to pass the time both for the gods and for the humans in Midgard. Games ranged from board games, to lawn games, to feats of strength and stamina, to all-out brawls that often ended in death (which I won't recommend), to swimming matches (or more accurately, drowning matches) where the goal was to hold your opponent underwater the longest (I don't recommend this either).[10]

KING'S TABLE

One popular board game mentioned often in the sagas is *hnefa-tafl* (King's Table). This game was very popular and was carried by the Vikings throughout their travels to alleviate boredom and strengthen their wits through strategy during long sea voyages and mead hall gaming sessions. We know little about the rules of the game or how the board was arranged, but thanks to references in the sagas and archaeological finds, we know that the game had pawn-like pieces (*hunns*, or "knobs") and a king (*nefi*) similar to chess.[11] Strategic thinking was an important

A modern replica of a *hnefa-tafl* game. Photo courtesy of Matěj Baťha, Wikimedia Commons.

skill to a Viking. Just as much as physical prowess, it could mean the difference between life and death. Because of this, strategic games—like poetry, dancing, and feats of strength and endurance—determined the status of a man among his peers.

In *Morkinskinna: The Earliest Icelandic Chronicle of the Norwegian Kings (1030–1157)*, brothers King Eysteinn and King Sigurd "Snake-in-the-Eye" (one of the four sons of Ragnar Lodbrok, the protagonist in the History Channel's popular series *Vikings*) are engaged in a test of wits in which they discuss their individual strengths. The brothers are participating in a feast and Sigurd is in a particularly dour mood, causing Sigurd's men to ask Eysteinn to approach his brother to determine why he has fallen silent. Thus begins the "Contest of Kings." At one point, Sigurd claims he is a "stronger and better swimmer," to which Eysteinn answers, "That is true, but I am more skilled and better at board games, and that is worth as much as your strength."[12]

DRINKING GAMES

Although the physical games were primarily for men, game playing was more than just a "sausage fest" (to quote my good friend Bobi Martin, who uses this term when referring to instances when us men-folk get together to drink, play games, and act like buffoons); women often participated in strategy games and drinking games. In the Icelandic saga *Egil's Saga*, we see an example of how women and men were paired up for drinking games by choosing lots, and usually stayed with the same partner until one or both of them passed out.[13] As described on the Games and Sports in the Viking Age page of Hurstwic, a website devoted to "Viking Combat Training," "the game consisted of pairs of men trading drinks and verbally sparring. With each drink, the participants were expected to compose and recite a verse of poetry, boosting their own reputation (with boasts of courageous and manly behavior) while disparaging their opponents (with taunts of cowardly or womanly behavior). As the drinking progressed, the intensity of the ridicule, boasts, and taunts increased as the drinkers became less and less inhibited. The goal was to maintain (or even enhance) verbal skill throughout the competition without showing the effects of alcohol."[14]

My friend Dave Brown (see his Odin's Golden Tears of Honey Joy Cinnamon-Vanilla Metheglin recipe on page 131) has developed his

Meads Worthy of Valhalla

A list of commercial meads compiled by Dave Brown to inspire your mead making and game playing. Each mead has been thoroughly quality-tested by myself and Dave.

Redstone Meadery Vanilla Beans and Cinnamon Sticks Mountain Honey Wine

Made from 2 parts alfalfa honey and 1 part wildflower honey, and fermented on whole vanilla beans and cinnamon sticks. Sweet but not overpoweringly so with hints of vanilla and cinnamon. A flavor party in your mouth, and the inspiration for the Odin's Golden Tears of Honey Joy Cinnamon-Vanilla Metheglin recipe. 12 percent ABV.

Valley Vineyards Sweet Golden Honey Mead

Extremely sweet and smooth. As close as you can get to drinking pure honey. Brewed in Ohio from fresh clover honey. 12.5 percent ABV.

Jaros Dwojniak Maliniak Polish Flavored Mead

Brewed in Tomaszów Mazowiecki, Poland. Marketed as "honey wine with raspberry juice," it has a sweet, thick honey flavor with a hint of raspberry, making for a heady sweetness with a touch of tart. Aged in oak barrels for five years. 16 percent ABV.

StoneBrook Winery Honey Mead

An affordable but very good dessert mead from Kentucky. 11 percent ABV.

Dansk Mjod Ribe Mjod

A "honey apple wine with hops added" from Denmark. Made from

own tabletop dice game called Don't Fall in the Mead Hall with the goal of emulating a Viking mead hall drinking game. The concept is simple, but has a subtle level of strategy that is heightened by any mead you happen to be drinking while playing it. To imagine what Dave's game is like, picture a group of Vikings sitting at their respective tables fresh, unfiltered, and unpasteurized apple juice. A touch of citrus and nice apple flavors with hints of cinnamon and brown sugar make for a very pleasant-tasting mead. The high alcohol content lends a nice warmth akin to a brandy. 19 percent ABV.

DANSK MJOD VIKINGERNES MJOD

A "Nordic honey wine" with hops. Has a fruity and sweet honey flavor. Lightly carbonated and more akin to a light, sweet beer (but not a bragot). Tastes better at room temperature than chilled but is good either way. 19 percent ABV.

DANSK MJOD VIKING BLOD

A "Nordic honey wine" with hibiscus and hops. The hibiscus gives it a bit of an earthy and almost spicy kick with just a touch of honey. Very unique and flavorful, but not for everyone. 19 percent ABV.

LURGASHALL TOWER OF LONDON MEAD

Sold as Malmesbury Whisky Mead in the UK, this is a traditional English mead fortified with Scotch for an extra bite and a hint of oak flavor. 20 percent ABV.

BUNRATTY MEADE

Imported from Ireland. This is not actually mead, but white grape wine with honey added after fermentation. Purists will pooh-pooh its existence on a mead list, but it is well worth mentioning. A strong honey wine with a taste similar to a hard liquor, it has a flavor that can be adjusted to your liking by mixing it with a "true" mead. That's how the Vikings would have done it. 15 percent ABV.

after the formalities of a *symbel* have been completed and the games and boasting have ensued. Naturally, the mead hall is well stocked with mead. As the Vikings imbibe enough to feel the effects of the horn's gift, they keep drinking. Each roll of the dice (or Odin's Breathalyzer) determines whether the Vikings at your table ask for another round of drinks, send a round of drinks to another table, or do a toast/*skål*, requiring that all tables drink.

Then there are the chairs. Depending on what you roll, your Viking may choose to pick up a chair and lob it at another table. Fighting ensues and your Viking either wins and goes back to his or her table, or is knocked unconscious and is effectively out of the game. The play continues until the Vikings at all tables but the winning one have been knocked unconscious, either by other Vikings or due to having had one mead horn too many. While the game is plenty of fun on its own, incorporating some of your homebrewed mead into the play adds an element of its own. Sound fun? Check out page 205 for more information.

Kubb

Kubb (sometimes called Viking Lawn Chess) is a popular game in Scandinavia. It's rumored to have been played in Viking times, but we don't know this for sure. My Danish friend Peter Androsof introduced me to it while he and his wife, Tina, were teaching Danish gymnastics for a semester at Berea College.

While you can purchase a set if you can find one, it's very easy to make your own (see the "Making and Playing Kubb" sidebar on page 188). My

The author's homemade kubb set displayed in front of his family's Appalachian Viking mead hall.

friend Tony Basham brought a set back from Denmark with tiny pieces that paled in comparison with the heavy-duty sets some friends and I made out of excess landscaping timber with wood dowels for throwing pieces. While it helps to follow the basic rules, don't feel the need to be tied down by dimensions. Just work with the space you have and make the pieces whatever size you prefer. Always be sure to enjoy some mead or ale while playing.

Tasting

While games, camaraderie, boasting, and even some skaldic meditation should all be part of your mead enjoyment, how do you drink the stuff? As we've learned, some meads are more like wines, some more like beers, and others refuse to fall into any category. While I do enjoy drinking mead from my drinking horn, or even straight from the bottle, there are more sophisticated ways to drink it that will allow you to experience the full mouthfeel and flavor. The first thing I do when I open a bottle of mead is put my ear to the bottle and listen. Even with a still mead you can sometimes hear the little creatures inside fizzing and popping with thankfulness for being unleashed upon the world once again. Then I pour some into a mead horn or a glass, swirl it around, and have another listen. Next, I take the first sip into my mouth slowly and swish it around a bit to see if the flavors I intended made it through, or determine (and hopefully appreciate) any unplanned flavors.

If I'm not in a swigging mood (a Viking would never set down a mead horn without first emptying it) and want to appreciate the clarity or coloring of the mead, I'll use a clear red-wine glass or a brandy snifter. These both have large openings that give plenty of space for air to mingle with the mead while allowing the aroma to waft through the opening. With a brandy snifter, the opening is slightly smaller than the bottom, allowing aroma and flavoring nuances to linger at the opening. For sparkling and dry meads, I'll use a tall, narrow white-wine glass. I will fully admit that I don't have the best sense of smell, so if I want to get a better idea of the overall qualities of a particular mead, I have my wife take a sniff and a sip and pass along her thoughts. There are more refined ways of going about this, but—perhaps due to my contrary nature and dislike of elitism—I tend to avoid the fancy-pants style of mead and wine tasting. I'm not saying people who like to sit back and

Making and Playing Kubb

As with my mead making, I like to stick within certain parameters, but don't feel the need to be tied down to conventions when it comes to kubb. I've provided the "official" dimensions and rules here, but I recommend not overthinking this. When playing, work with the space you have, and make the pieces whatever size you prefer if creating your own set. That's how the Vikings would have done it.

Materials

6 feet (1.8 m) of 4×4 lumber or landscape timber
6 feet (1.8 m) of 1½- to 2-inch (3.8- to 5-cm) wood dowel rod for batons (softwood makes for a more challenging throw)
6 feet (1.8 m) of ¾-inch (19 mm) wood dowel rod for corner stakes
30 feet (9 m) of string

Process

1. To make the king, cut 12 inches (30 cm) from the 4×4 lumber.
2. Rip (cut down the middle) the remainder of the 4×4 lumber to 2¾ × 2¾ inches (7 × 7 cm).
3. Cut the remaining wood to create ten 6-inch (15-cm) kubb pieces.
4. As an alternative to ripping, you can make a monster set like the one I made from old landscaping timber. My kubbs are 12 × 3 inches (30 × 7.5 cm), and my king is 6 × 16 inches (15 × 40 cm) made from two 16-inch-long pieces glued together.
5. Decorate the king and pawns in any manner you wish. Some people will cut a "crown" into the king; for mine, I took a wood burner and burned runes into each piece.
6. Cut the ¾-inch (19 mm) dowels to six 12-inch (30 cm) pieces; miter (cut at an angle) the end of each to create a stake.
7. Cut six 1¼-inch (3.2 cm) dowels by length to 12 inches (30 cm) each.

Play Some Kubb!

The rules are simple, making first-time players somewhat skeptical about just how addictively fun it can be.

1. The first step, of course, is to make sure each player has some mead to drink.
2. Next, stake the playing field by measuring out a 16 × 26-foot (5 × 8 m) rectangle. Place four batons at each corner and one in the direct center of each 26-foot (8-m) line.
3. Set the king in the center of the field between the center stakes and space five kubbs evenly along each baseline (the 16-foot [5-m] section).
4. Each team (you can play this one-on-one or in groups of two or three) takes to their baseline and throws all six batons at their opponent's kubbs. Batons must be thrown underhanded and end-over-end (no horizontal spinning). They may rotate vertically.
5. If the first team (Team A) knocks over any kubbs, a player from the opposing team (Team B) must pick up each fallen kubb at the beginning of their turn, stand behind their baseline, and toss the kubb somewhere onto Team A's side of the field (the other side of the centerline). These "field kubbs" must then be stood up where they land. One mulligan (do-over) is permitted if the kubb falls outside the baseline. If the second attempt doesn't fall within the baseline, Team A can place it within one baton's length of the centerline.
6. The next team must first knock over any field kubbs before attempting to knock over baseline kubbs. Any baseline kubbs mistakenly knocked over should be stood up.
7. If field kubbs are still standing at the beginning of the *next* round, the opponent then gets to use the kubb nearest the king as their new baseline.
8. When a team has knocked over all its opponent's kubbs, they then have one chance each round to knock over the king (from behind the baseline).
9. When the king topples, the game ends. If any team mistakenly topples the king while kubbs are still standing on their opponent's side, they immediately lose. The other team reserves the right to mock them mercilessly.
10. Since games tend to be short, it is common to play best-out-of-three matches.

A glass of a clear, well-aged mead. The large opening of the wine glass allows for plenty of air contact to impart the full effect of the aroma.

enjoy the bouquet of a good glass of wine or mead are necessarily elitist; it's just not the sort of thing that the Viking in me is into.

I prefer to enjoy all my meads and beers at cellar temperature. I will chill some dry, sparkling meads, cysers, melomels, and light beers and bragots a bit more, but most times I open a bottle shortly after I've taken it from the cellar. I never chill my glasses, even in the summer. I used to do this, but after learning that drinking from a cold glass detracts your taste buds' ability to identify flavor, I have stopped doing so.

Mead Circles

More and more often, I hear about people getting together for mead circles or "meadings" to pass around bottles of homemade mead, talk about the personality of their mead and the story behind how they made it, and maybe pick up a buzz unlike any they've ever had. After sipping from several bottles, you may pick up a bit of a boozy buzz, but the biggest rush you'll get is from the array of flavors lingering on your tongue and the rush of sharing handcrafted meads with a like-minded crowd.

As these meadings progress, ideas are discussed and games are played. One of these games is "pirate mead," in which a little mead from each participant's batch is poured into a separate bottle until it's full. The bottle is then capped, buried, and dug up at the next meading to be passed around. I encourage you to ask around for mead circles in your area (online sources for helping locate one can be found in this book's resources section). They generally aren't publicly advertised, as organizers prefer to keep out the sort of folks who are just interested in getting drunk and have no homebrew of their own to share. This isn't to say that they won't allow someone to join who hasn't brewed anything yet if that person shows genuine interest. Mead makers generally love to introduce others to the joy of mead making and will likely invite you back to share one of your own brews. If you are unable to locate any mead circles in your area, create your own. You can start by simply getting together with one or two other mead makers you know, or sharing your mead with friends who then may show an interest in learning how to do it themselves. Before you know it you'll be a thriving group of passionate mead makers sharing a gamut of meads, bragots, and grogs with all manner of ingredients. Above all else, have fun. You're a Viking—you can do this!

Closing Thoughts

More and more people are starting to take part in the underground wild-mead movement. This movement isn't just about making your own booze, it's a crusade to forge a web (or perhaps a hive?) of like-minded individuals who want to approach all aspects of their lives holistically and naturally. It's a meeting of minds who are passionate about fermentation and who wish to befriend the bacteria that are quite literally responsible for who we are as individuals. In the end, it's about wanting to have a good time while living an authentic life.

As someone who is a direct product of the back-to-the-land movement of the 1970s, I was raised with the ethos of doing things for myself and using what the land has to offer whenever possible rather than relying primarily on monetary transactions. For me, and others like me,

VÖLUSPÁ

At the beginning of this book, I referenced a section of *Völuspá* from *The Elder Edda*. I would like to close with the ending of this poem, which is as relevant now as when it was it was passed along by skalds and scribes in ancient times. While reading it, think on how the Norse viewed time as cyclical. They may have viewed humankind—and the gods—as doomed from the beginning, but when Ragnarok comes and ends the world as we know it, the earth will heal itself and be green once again. Like the gods and their unceasing battle to avert predestined doom and destruction, let us each do our part—despite seemingly insurmountable odds—to keep fighting against those who seem bent on destroying the wildness of our Midgard.

> Garm is howling from the Gnipa Cave,
> the rope will break, and the Wolf run free.
> Great is my knowledge, I can see
> the doom that awaits almighty gods.
>
> . . . .
>
> Mimir's sons play; now fate will summon
> from its long sleep the Gjallarhorn:
> Heimdall's horn clamors to heaven,
> Mimir's head speaks tidings to Odin.
>
> Lofty Yggdrasil, the Ash Tree, trembles,
> ancient wood groaning, the giant goes free;
> terror harrows all of Hel,
> until Surt's kinsman comes to consume it.
>
> . . . .
>
> The sun turns black, the earth sinks below the sea,

no bright star now shines from the heavens;
flames leap the length of the World Tree,
fire strikes against the very sky.

She sees the earth rising again
out of the waters, green once more;
an eagle flies over rushing waterfalls,
hunting for fish from the craggy heights.

. . . .

Barren fields will bear again,
Balder's return brings an end to sorrow.
Hod and Balder will live in Odin's hall,
home of the war-gods. Seek you wisdom still?

She sees a hall, fairer than the sun,
thatched with gold; it stands at Gimlé.
There shall deserving people dwell
to the end of time and enjoy their happiness.

There comes the dark dragon flying,
flashing upward from Nidafells;
on wide swift wings it soars above the earth,
carrying corpses. Now she will sink down.

—from *Völuspá*, translated by Patricia Terry[15]

As Patricia Terry, the translator of this version of the *Völuspá*, wrote about the last stanza, "[It has been] the subject of much conflicting interpretation, in which the dragon is seen in a variety of functions from purifying to threatening . . . I see its presence as a reminder that good cannot be disentangled from evil; to separate light from the darkness is to intensify the darkness."[16]

this way of living has become second nature. In the last 10 to 20 years, a new generation of back-to-the-landers has arisen. Descriptive terms I've heard attributed to those who are part of this subculture include *DIY*, *green living*, *sustainable living*, and *holistic living*. While I think these are all great descriptors and encourage their use, I prefer to think of this lifestyle as simply "living" (or "living simply"), as this is what it was called in the days of our ancestors. They lived this way because they had to. It wasn't always easy, nor was it always fun, but it was necessary, authentic, and encouraged a mind-set that resulted in hardy individuals who could live through the toughest of times using what little they were able to produce on their own—and by working with their neighbors and family members to share their resources, ideas, and passions.

My hope is that readers of this book will continue this tradition in their own communities. Get to know your local beekeepers, start keeping your own bees, and combine the bee's gift with bounty from your garden and wild lands—and water from local springs—to create and share healthful, ecstasy-inducing fermented beverages that are uniquely *you*. Who knows where this journey will lead you. The paths an addiction to fermentation can open are endless, from healthy lacto-fermented vegetables, to probiotic kefir and yogurt, to cheese and salami, to moonshine distilled from excess mead and wine (and enjoyed in moderation, of course).

We can extend this concept into our fermentation endeavors. In making our own healthful and magical fermented wild beverages, why not take the time to talk—and listen—to the billions of microbes, plants, and other living beings that live among us? If we're willing to learn their language, perhaps we can come to the realization that there are no "good" or "bad" bacteria, but rather that all our little friends have something to offer if we just give them the chance. If the Vikings could integrate with cultures that initially saw them as foreign—and thus frightening—we can learn from them and embrace a global culture of communal fermentation.

You're a Viking—you can do this! *Skål!*

✤ ACKNOWLEDGMENTS ✤

The people I would like to acknowledge for helping this book come to fruition are too many to name here. Just as it takes a village to raise a child, it took a community to write this book. From the people I have met at my workshops, to the passionate fermenters and sustainable-living devotees I have engaged with online and at skill-sharing festivals, to the many writers and thinkers who have touched my soul with their words and inspired me through the best and worst of times, to my family and friends—I offer a horn of mead. In particular, I would like to thank (nonexclusive and in no particular order): The teachers who have inspired my love of writing and reading from an early age, starting with my father, Wayne, who was known by his students as "Mr. Z" while a speech, drama, and English teacher at Lloyd High School in Erlanger, Kentucky, and the professors at Berea College who were integral in helping me find my voice and fine-tune my writing, particularly John Bolin and Richard Sears. The list of writers whose material has both inspired and educated me is far too long: Terry Pratchett, Douglas Adams, and Neil Gaiman for their wit, wisdom, and storytelling; Sandor Katz, Stephen Harrod Buhner, Patrick McGovern, and Janisse Ray for their instructive and insightful nonfiction writing; my ancient Northern European and Anglo-Saxon ancestors for the myths and stories they passed down; and the editors, folklorists, and philologists who didn't allow these stories to die. And then there are those I have already thanked in some capacity: my good friend David "Dave" Brown—who first greeted me at five years old dressed as Superman and whose vast and rather odd imagination helped fill countless childhood (and adulthood) hours; Zach Zimmerman for building me a Viking mead hall on our parents' farm (and almost cutting his foot off with a chain saw in the process); and the third member of our mead- and Viking-obsessed crew, Steven "Stickboy" Cole: May your fanciful writing and illustrations someday make it into the hall of literary oddities (I look forward to sharing a place with you there). Every single member of the Earthineer community, particularly Dan Adams for his vision to create an online haven for homesteaders and homestead dreamers, his wife (and my cousin) Leah, ShadyGlade Farm, GrumpyOldMan, GnomeNose, BickensChickens, Spun Gold Farm, Jimbo, Jason, Bearclover, and the many other members whose names and online handles I can't fit here. I took my first faltering steps writing about mead for Earthineer and the encouragement, sarcasm, and conversation each of you offered were in a very real way a motivation for this book. And finally, the staff at Chelsea Green, in particular my editor Michael Metivier, for taking a chance on my particular brand of oddness. Michael has been a gracious and insightful editor. I feel I've gained a friend in the process.

I offer my eternal gratitude and a rousing *skål* to y'all!

☙ TROUBLESHOOTING ☙

L ike it or not, no matter how much attention you pay to your ferments, problems will arise. Sometimes you get unwanted microbes in your brew, or have unexpected results due to extraneous factors such as seasonal temperature fluctuations. Don't let this get you down. Even though it's hard not to get upset when a ferment you've invested so much time and love into doesn't turn out quite right, remember that the unique flavors of the ferments we enjoy today had their start in fermentations that went awry or were unintentional to begin with. While I sometimes learn to simply enjoy the uniqueness of a ferment that didn't go quite as planned, there are techniques you can employ to resolve or avert problems.

FERMENT IS SLOW TO START

A slow or weak fermentation can be caused by a variety of factors depending on what you're fermenting and your method of introducing yeast. Mostly, it will come down to three things: aeration, temperature, and amount of nutrients and sugars present.

Anything with a lot of honey in it can take a few days to a week to show active fermentation due to the amount of sugars. Oxygen helps move this process along, which is why starting a ferment in an open-mouth container and stirring several times a day is a good trick for introducing the *bryggjemann*, whether you're wild-fermenting or adding yeast cultures. Other factors, such as introducing a strong yeast or active barm, or including less honey or sugar in the initial ferment than what you plan on having in the final product, can help get a fermentation started quicker.

Try to always initiate a fermentation in a warm area, preferably no cooler than 70° F (21° C). If fermentation still seems slow to start, increase the heat in your brewing room, move the vessel to a warmer area of the house, or transfer the must to an airlocked container and wait for warmer weather. In combination with the first two options, it doesn't hurt to add botanicals for additional nutrients, acids, and tannins to give the yeast something to feed on. Add only small amounts (a few teaspoons, ounces, or grams), and strain them out when you rack to prevent an overly astringent or acidic flavor in the final product. Always make sure you use water that is dechlorinated and contains no chloramines (see chapter 4). If you haven't, you may need to add more yeast mixed in with warm water that meets these requirements.

FERMENTATION STARTED, BUT IS NOW "STUCK"

All of the tips referenced for stimulating a slow or weak fermentation will work for restarting a stuck fermentation. You can usually tell a fermentation is stuck if it's overly sweet and there is little to no activity in the airlock. Although you can test the alcohol level with a hydrometer, usually tasting a bit will indicate whether or not the alcohol level is too low.

Generally, I judge whether or not fermentation is moving along as it should by watching the carboy regularly and noting when I first started the fermentation. If the mead is fairly clear after a month or two and there is an inch or more of lees on the bottom, the fermentation is moving along just fine. Either way, I almost always rack after a couple of months.

If the fermentation is slow or stuck, be sure to swirl the carboy around as you're racking. You can also pour it into an open fermenter and stir it with a totem stick for at least five minutes. If you don't want to go to the effort of racking all the mead, siphon out a couple of cups into a small container and stir or shake vigorously, or blend in a blender on a high setting.

Sometimes you will simply have provided too much sugar for a full fermentation. Siphoning some of the mead into a jug with an airlock and topping off the carboy with water might help this. Other times the fermentation will have started but slowed down because the yeast ran out of sugar to consume, or the sugar didn't mingle well with the yeast. This is a case in which aeration should help.

If you're brewing with wild yeast, you may have picked up a weak yeast strain. This is where it doesn't hurt to add a high-gravity commercial yeast or some barm from an active batch of recently brewed mead.

As with all ferments, regular attention is a good policy. Whether it's communing with your mead regularly, taking careful notes, or invoking the brew gods, if you show your mead love it will return the favor. With all of the above in mind, remember not to rush the process. I've started meads in the cold of winter that were sluggish to begin with but picked up as the weather warmed. Sometimes aeration or sugar adjustment played a role, but other times it simply started up on its own when it was good and ready.

Mead Is Too Tart, Harsh, or Insipid

This can be resolved through acid or tannin adjustments, as discussed on pages 61 thru 64.

Mead Is Too Sweet, Dry, Tart, or Sour Prior to Bottling

If you're confident your mead has fully fermented and aged in the carboy long enough for bottling, give it a taste before bottling. This is a good time to add acids or tannins as discussed on pages 61 thru 64. You may find on tasting that it seems just a bit sweeter or drier than you were hoping for, or perhaps a bit too tart or sour. As noted in On Sourness and Funkiness in Mead on page 86, these flavors can dissipate with aging, so don't fret too much over it.

Sweetening is simple, and will help dissipate tart and sour flavors if done properly. Just add a cup of honey-water, stir, and taste, and repeat until you're satisfied. This is called "back-sweetening." You may want to leave it in the carboy a few weeks or even a few months longer, as the honey may initiate a small amount of fermentation, which could at best lead to a sparkling mead if bottled now, or at worst a mead grenade. Stirring slowly while adding sweetener will lessen the chances of restarting fermentation. However, you will always have some degree of yeast remaining in any mead. Using a yeast with high alcohol tolerance and fermenting to bone-dryness before sweetening is one way to minimize the chances of restarting fermentation, but this will also result in a high-alcohol "rocket fuel." Lactose, or milk sugar, can also be used. This is a complex sugar found in milk that is essentially dehydrated whey. A traditional way to cut off fermentation to retain some residual sweetness was to add a bit (½ cup or so) of a high-alcohol, taste-neutral liquor such as vodka to kill off any remaining yeast. If you're not entirely opposed to adding chemicals, you can add ½ teaspoon of potassium sorbate per gallon before adding additional honey. This keeps the yeast alive while not allowing fermentation to continue. The powdered form of potassium sorbate available at homebrew stores is a chemical clone of the natural sorbic acid molecule, which was first discovered in the berries of the mountain ash tree. You very likely ingest it on a regular basis, as it is one of the most widely used food preservatives. Because of this, it has been tested extensively, and has been determined to be non-carcinogenic and nonmutagenic.[1] It's also an approved chemical in the *Handbook of Green Chemicals*.[2] Still, I always take studies undertaken on any chemical used by the food industry with a grain of salt. I recommend patience, instinct, and offerings to Odin as an alternative to chemicals. Learn to enjoy the variety of flavors you produce (even the unexpected ones) and become accustomed to sharing your mead with friends who have different taste preferences.

Drying out a mead is best avoided unless you feel it is far too sweet. The best way to do this is to use the some of the methods you would for a stuck fermentation to restart fermentation. The most reliable method is to add a champagne or other high-gravity yeast. This may cause the mead to become too dry, or throw off the flavor balance. You'll then need to make additional flavoring adjustments before bottling. You can also add more water—but do this in small doses, or you could end up with a watered-down final product. Unless you're particularly picky about the flavor and alcohol level of your mead, I'd suggest avoiding going to all this effort. But then again, I'm a lazy Viking—or rather, a busy home-steader—and prefer a low-maintenance approach to mead making.

❧ BREWING GLOSSARY ❧

Acerglin: Mead made with maple syrup.

Acetic acid: The acid that is produced when a fermentation turns to vinegar. Not to be confused with citric, malic, and tartaric acids, which in abundance cause tartness, but can help balance the flavor of a mead otherwise.

Acetobacter: An ubiquitous airborne bacteria that can cause any alcohol ferment to metabolize into acetic acid given enough contact with oxygen.

Acid blend: Citric, malic, and tartaric acids that are blended and sold in home-brew stores to mimic the natural acids found in grapes and many other fruits and botanicals.

Adjunct: Ingredients added to a mead, wine, or beer that aren't absolutely necessary, but will aid fermentation or provide a more complex flavor profile.

Aeration: The process of stirring the liquid in a vessel, swirling the vessel, or racking into another vessel to incorporate large amounts of oxygen. Necessary for initiating a wild fermentation or ensuring an active fermentation once fermentation commences. Also a technique for restarting a "stuck" fermentation.

Aerobic bacteria: Bacteria (such as acetobacter) that require oxygen to survive.

Airlock: A simple mechanism that allows carbon dioxide to escape a fermentation vessel while preventing outside air or bugs from entering.

Alcohol by volume (ABV): Percentage of alcohol per volume in mead, beer, or wine.

Ale/beer/cider: Terms that are often used interchangeably in historical references to brewing. Today *ale* generally refers to a beer that is top-fermented at warmer temperatures than a lager (cold- or bottom-fermented beer). To make it more complicated, the words *ale*, *beer*, *cider*, and *mead* can often mean the same thing when referenced in literary or historical sources, and the word that *beer* stems from (*beor*) very likely may have originally been in reference to cider.

All-grain: Beer made from only grains rather than from extract. A relatively new term, as all beers were made this way prior to the modern homebrewing movement.

Anaerobic bacteria: Bacteria that do not require oxygen to survive.

Astringent: A dry, puckering mouthfeel that is a result of too many tannins present in a mead. In small amounts, can help give a mead body.

Attenuation: Full attenuation means all fermentable sugars have been converted to alcohol and carbon dioxide. If a mead is not fully attenuated, this means that there is residual sugar that will cause sweetness, or potentially further fermentation.

Backslopping: The process of initiating fermentation by incorporating a small amount of a previous active ferment (also known as *barm, bug,* or *starter*) into a new batch.

Back-sweetening: Sweetening a "dry" batch of mead (or beer, wine or cider) through the addition of unfermentable sugar, or by adding fermentable sugar and chemically halting the fermentation through the addition of sulfites.

Barm: A word often seen in older cookbooks and brewing texts to reference the incorporation of an active yeast (usually liquid but sometimes dry) into a newly created batch of must or wort to initiate fermentation. See *backslopping.*

Beer: See *Ale/beer/cider.*

Bottle-conditioning: Bottling beer or mead before fermentation is complete to allow secondary fermentation in the bottle, creating complex aromas and flavors, and usually a degree of carbonation.

Bragot (also *braggot, brag, bragio, brakkatt,* or *bracket*): A mead-beer hybrid made with honey and malt. It was traditionally flavored with herbs and spices, and occasionally with hops.

Bug: Similar to *barm* or *backslop* in that a bug is an active ferment used to initiate future ferments. A bug is created intentionally as a starter using a small amount of liquid rather than being taken from a larger volume of an active ferment.

Capsicumel: Mead made with chili peppers.

Carbonation: Carbon dioxide that is created by adding sugar or honey to a fermented liquid prior to bottling to initiate bubbling or "sparkle." Can also be created by cutting an active ferment off completely from outside air contact. Potentially dangerous if proper precautions aren't taken.

Cider: See *Ale/beer/cider.*

Cyser: Mead made with apple or pear cider.

Fortified mead: Mead with distilled liquor added to it to increase alcohol content and depth of flavor. Fortifying is also a traditional technique for long-term preservation of mead.

Gesho (*Rhamnus prinoides*): A species of buckthorn native to eastern and southern Africa. Its sticks and leaves are used in making t'ej (Ethiopian honey wine).

Hulled/unhulled seeds: Unhulled grain seeds are required for sprouting/malting/germination of grains for grain-based ferments (beers, ales, bragots, et cetera). Hulled seeds are those that have had the outer layers removed.

Lactomel: Mead made with milk.

Lees: Dead yeast cells that fall to the bottom of the brewing vessel as yeast devours sugar and enzymes to produce alcohol in mead or wine. For beer, it is called trub. Results in a sediment from which the liquid must be racked.

Malt: Barley or other grains that have been germinated or sprouted, releasing enzymes that break down complex carbohydrates into simple carbohydrates, which can then be fermented into alcohol.

Malt extract: Liquid wort that has been condensed into a syrup or powder for reconstitution by homebrewers. The process requires less time and equipment than all-grain brewing, but is also less flexible in regard to recipe fine-tuning.

Mashing: In beer and bragot brewing, the process of immersing crushed grains and adjuncts in hot water to convert grain starches to fermentable sugars and non-fermentable carbohydrates, producing a sugar-rich liquid called wort.

Mead: See this book.

Meade: A white grape wine that has been blended with honey.

Melomel: Mead made with fruit.

Metheglin: Mead made with herbs and spices.

Morat: Mead made with mulberries.

Mouthfeel: The sensation of consistency or viscosity when tasting an alcoholic beverage.

Must: Unfermented wine or mead, or "new wine." Originally an Old English word stemming from the Latin *mustum*, shortened from *vinum mustum* (fresh wine), and the neuter of *mustus* (fresh, new, newborn).

Odroerir: Stems from *Óðrerir*, an Old Norse word for "inspirer of wisdom," and is often used in reference to poetry. It can also refer to mead, or the vessel in which mead is brewed (or served). An (almost) literal translation is "mead of poetry."

Omphacomel: Mead blended with verjuice (the acidic juice of unripe grapes).

Open fermentation: Sometimes confused with wild fermentation, the process of fermenting in a large-mouth vessel covered by a porous cloth to provide aeration. Can be done with both wild and commercial yeast fermentations.

Oxymel: Mead blended with wine vinegar.

Pasteurization: Sterilization through heating. Commonly applied in modern times to remove most or all bacteria from honey, milk, or other substances. In mead making, pasteurized honey (or must heated to pasteurization temperatures) can remove bacteria with positive health and flavor benefits.

Probiotics: Bacteria (or friendly pixies?) present in many natural ferments that bestow health benefits to the organism that ingests them, creating a symbiotic relationship of sorts.

Pyment: Mead made with grapes.

Racking: Transferring (by siphoning or pouring through a funnel and strainer) an incomplete alcoholic ferment into another vessel, leaving behind lees (yeast sediment) that can cause cloudiness or a yeasty flavor in the final ferment. Can also be used to aerate or restart a "stuck" fermentation.

Reinheitsgebot: The purity law instituted in Bavaria in 1516 that subsequently was extended to all German brewers making beer that was to be consumed in Germany. The law requires the use of malted grains, hops, yeast, and water only.

Rhodomel: Mead made with attar, an essential oil distilled from rose petals.

Rhyzamel: Mead made with root vegetables.

Sanitization: The use of natural or man-made chemicals to minimize bacteria present on the surfaces of equipment used in the brewing process. Often confused with sterilization.

Sima: A traditional Finnish mead made with lemons to celebrate May Day. Most modern recipes call for all or most of the honey to be replaced with sugar.

Skål (or skol): Although the literal translation from Old Norse and modern Icelandic is "bowl," this is a common Scandinavian toast that goes back to the time of the Vikings.

Sparging: In beer and bragot brewing, the process of pouring hot water over spent mash grains to retrieve malt sugar and extract from grain husks.

Starter: A small amount of an actively fermenting liquid used to initiate fermentation.

Sterilization: The use of high temperatures to eliminate all bacteria from the surfaces of equipment used in the brewing process, as opposed to sanitization, which only eliminates most bacteria.

Tannins: Bitter or astringent chemical compounds present in many botanicals that are important for adding body to an alcoholic ferment. Also available in powdered form at homebrew-supply stores. Best in moderation.

T'ej: The national drink of Ethiopia. Traditionally made from wild-fermented honey-water and gesho (for bittering and wild yeasts).

Trub: See *lees*.

Wild fermentation: A process for initiating a ferment solely through use of wild yeasts present in the air, or in a substrate such as organic botanicals or raw honey.

Wort: Unfermented beer; literally "plant," as beer was traditionally made with not only harvested and malted grains, but also various herbs such as mugwort and Saint-John's-wort. In Old Norse it is *rot* (root).

Yeast: Eukaryotic microorganisms that are members of the fungi kingdom, and are agents for fermentation.

Zymurgy: The scientific study of fermentation. A branch of applied chemistry, it is most often used in reference to brewing. Also the name of the official magazine of the American Homebrewers Association (AHA).

❊ RESOURCES AND INSPIRATION ❊

WEBSITES

www.earthineer.com: A "peer-to-peer social marketplace for food and farm" with articles, discussions, and bartering groups on brewing, fermentation, and plant use. Also a repository for many of the author's articles on mead making and other subjects.

www.jereme-zimmerman.com: The author's personal website, with information and regular updates on his writing and workshops. Information on Dave Brown's tabletop dice game *Don't Fall in the Mead Hall* can also be found here, or at vikingnerds.com.

www.truesourcehoney.com: An organization of honey producers who voluntarily place the TRUE SOURCE HONEY label on their honey jars to demonstrate their commitment to producing "real" honey. An independent third party monitors the sourcing practices of members through the True Source Certified voluntary traceability system to ensure compliance with US and international trade laws, enabling the honey to be tracked to its country of origin and even the beekeeper who harvested it.

www.bushfarms.com/bees.htm: The companion website to beekeeper Michael Bush's book *The Practical Beekeeper: Beekeeping Naturally* (X-Star Publishing Company, 2011).

www.medievalcookery.com: Lots of medieval mead, wine, and beer recipes, as well as plenty of food recipes for preparing a feast worthy of a Viking.

brundo.com: An online store of Ethiopian herbs and spices, including gesho for making t'ej.

www.gotmead.com: "Your mead resource" is just that. An impressive database of mead-making resources and forums on all manner of mead-making styles.

www.vikinganswerlady.com: A very thorough and informative source for questions on Viking history, Norse mythology, and daily life of early Scandinavian cultures. Some of my earliest forays into brewing like a Viking were inspired by the Old Norse Alcoholic Beverages and Drinking Customs section.

norse-mythology.org: Also known as "Norse Mythology for Smart People: The Ultimate Online Resource for Norse Mythology and Religion." A product of Dan McCoy, this site provides thorough, well-documented descriptions of key points of Norse and Germanic mythology with a touch of philosophical pondering.

www.wildfermentation.com: An invaluable resource for information on fermenting all manner of food and beverages with several forums, this is the companion site to Sandor Katz's book *Wild Fermentation* and his follow-up *The Art of Fermentation*. I can't overestimate the impact of Katz's work on my fermentation practices.

www.meadmadecomplicated.org: The name says it all. If you thought this book wasn't complicated enough, go here. Really, though, it does have some great information for those interested in the technical details of mead making.

www.mazercup.com: The website for the Mazer Cup International Mead Competition in Broomfield, Colorado.

mead.meetup.com: A resource for finding mead, brewing, and fermentation groups to "mead" up with.

www.pitt.edu/~kloman/tej.html: An in-depth resource on t'ej (Ethiopian honey wine).

www.homebrewersassociation.org: The official website of the American Homebrewers Association (AHA) and its magazine, *Zymurgy*.

www.spruceontap.com: Offers spruce tips harvested sustainably from trees in the San Juan Mountain Range of southern Colorado, as well as wild-harvested ingredients common to traditional Scandinavian and early American meads and ales such as yarrow and juniper berries.

www.plantsandhealers.org: The website for Plants and Healers International (PHI), a nonprofit organization devoted to archiving and spreading the lifework of the late ethnobotanist Frank Cook. PHI "facilitates knowledge exchange, skill building, inspires individuals and equips communities with tools to transition to a sustainable society that is in harmony with the natural world."

www.botanyeveryday.com: Another site dedicated to passing along Frank Cook's vast knowledge in plant lore and global connections by "evolving a multi-spectrum understanding of the plant human interface from varied perspectives." Ethnobotanist Marc Williams teaches classes in the identification and use of plants for food and healing through online courses offered through the website, carrying on the tradition of Frank Cook's donation-only email course.

sustainablekentucky.com: Writer Jamie Aramini documents her passion for sustainable living in Kentucky by profiling farms, farmers, and foragers. Jesse Frost's simple wild-fermented mead recipe on this site shows just how easy mead making can be.

www.roughdraftfarmstead.com: Jesse and Hannah Frost's adventures in off-grid homesteading are documented here, including several recipes for wildcrafted and fermented foods and beverages.

botanical.com: A great resource for researching herbs and their traditional usage in medicine and brewing. Essentially a hypertext version of the 1931 book *A Modern Herbal* by Mrs. M. Grieve, a vast resource in itself on the usage, folklore, and history of herbs.

www.reclaimingyourroots.com: A website and newsletter based out of Knoxville, Tennessee, run by herbalist, herb grower, and wellness educator Rachel Milford.

www.academia.edu: A good resource for finding current academic research into ancient fermentation practices.

www.sacred-texts.com: "The largest freely available archive of online books about religion, mythology, folklore and the esoteric on the Internet." An excellent source for public-domain Icelandic and Norse mythology texts.

www.pixiespocket.com: A thorough and thoroughly enjoyable website devoted to wild-foraging and natural living, with several one-gallon mead and wine recipes. The Head Pixie, Amber, can be seen frolicking playfully all over the Internet.

www.alehorn.com: AleHorn is a "drinking horn company that produces fine handcrafted items for modern barbarians." In addition to their high-quality horns, tankards, and other drinking implements, they offer horn beard combs, other mead- and Viking-related accessories, and an entertaining blog.

www.meadcrafter.com: A repository of mead resources, including recipes, an online shop, a blog with discussions on technique and ingredients, and a meadery directory.

BOOKS

Ehle, John. *The Cheeses and Wines of England and France.* South Deerfield, MA: New England Cheesemaking Supply, 1972.

Enright, Michael J. *Lady with a Mead Cup: Ritual, Prophecy and Lordship in the European Warband from La Tène to the Viking Age.* Dublin: Four Courts Press, 1996.

Fitzhugh, William F., Elisabeth Ward, eds. *Vikings: The North Atlantic Saga.* Washington, DC: Smithsonian Books, 2000.

Frost, Jesse. *Bringing Wine Home.* Bugtussle, KY: Rough Draft Farmstead, 2013.

Guerber, H. A. *Myths of the Norsemen: From the Eddas and Sagas.* New York: Dover Publications, 1992. Reprint, London: George G. Harrap and Company, 1909.

Jolicoeur, Claude. *The New Cider Maker's Handbook: A Comprehensive Guide for Craft Producers.* White River Junction, VT: Chelsea Green Publishing, 2013.

Katz, Sandor Ellix. *Wild Fermentation: The Flavor, Nutrition, and Craft of Live-Culture Foods.* White River Junction, VT: Chelsea Green Publishing, 2003.

Logsdon, Gene. *Good Spirits: A New Look at Ol' Demon Alcohol.* White River Junction, VT: Chelsea Green Publishing, 2000, 1–2.

McGovern, Patrick E. *Uncorking the Past: The Quest for Wine, Beer, and Other Alcoholic Beverages.* Berkeley: University of California Press, 2009.

Metzner, Ralph. *The Well of Remembrance: Rediscovering the Earth Wisdom Myths of Northern Europe.* Boston: Shambhala, 1994.

Mosher, Randy. *Radical Brewing: Recipes, Tales, and World-Altering Meditations in a Glass.* Boulder: Brewers Publications, 2004.

Nordland, Odd. *Brewing and Beer Traditions in Norway: The Social Anthropological Background of the Brewing Industry.* Oslo: Universitetsforlaget, 1969.

Papazian, Charlie. *The New Complete Joy of Home Brewing.* New York: Avon Books, 1991.

Savage, Anne, trans. and comp. *The Anglo-Saxon Chronicles.* New York: St. Martin's/Marek, 1983.

Schramm, Ken. *The Complete Meadmaker: Home Production of Honey Wine from your First Batch to Award-Winning Fruit and Herb Variations.* Boulder: Brewers Publications, 2003.

Sparrow, Jeff. *Wild Brews: Beer Beyond the Influence of Brewer's Yeast.* Boulder: Brewers Publications, 2005.

Stewart, Amy. *The Drunken Botanist: The Plants That Create the World's Great Drinks.* 4th ed. Chapel Hill, NC: Algonquin Books, 2013.

Terry, Patricia, trans. *Poems of the Elder Edda.* Philadelphia: University of Pennsylvania Press, 1990.

FESTIVALS AND EVENTS

Berea Festival of Learnshops: An intensive hands-on learning festival held for two weeks in July offering classes ranging from two hours to three days by a diverse range of artisans and educators. Held in the author's hometown of Berea, Kentucky; mead-making and fermentation courses are also offered.

The Firefly Gathering: A skill-sharing gathering that lasts several days and is held late June–early July near Asheville, North Carolina. Often has multiple mead-making, fermentation, and wild edible/medicinal plant workshops

Florida Earthskills Gathering: Similar to the Firefly Gathering. Held in Hawthorne, Florida, in early February.

The Mazer Cup: An international mead competition held in Broomfield, Colorado, in March. Participants can mail or drop off entries.

Moonshadow Food for Life Conference: An annual conference held at the Sequatchie Valley Institute (SVI) near Chattanooga, Tennessee, usually in early spring (dates vary). Offers "experiential learning opportunities" in areas such as holistic food preparation and preservation, fermentation, and natural beer/wine/mead brewing.

The Whippoorwill Festival: A skill-share gathering held the second weekend of July near Berea, Kentucky.

❋ NOTES ❋

CHAPTER ONE: The Mythological Origins of the Magic Mead of Poetry

1. H. A. Guerber, *Myths of the Norsemen from the Eddas and Sagas* (New York: Dover Publications, 1992; reprint of original publication by George G. Harrap and Company, London, 1909), 1.
2. Patricia Terry, transl., *Poems of the Elder Edda* (Philadelphia: University of Pennsylvania Press, 1990), 1–3.
3. Dan McCoy, *Norse Mythology for Smart People: The Ultimate Online Resource for Norse Mythology and Religion*, http://norse-mythology.org/kvasir.
4. Ibid.
5. Guerber, *Myths of the Norsemen*, 96.
6. Ibid., 96–99.

CHAPTER TWO: Mead in the Viking Age

1. *The Mabinogion*, trans. Lady C. Guest, "Everyman" edition, p. 278. Quoted in Hilda M. Ransome's *The Sacred Bee in Ancient Times and Folklore*, (Mineola, NY: Dover Publications, Inc., 2004; originally published in 1937 by George Allen & Unwin, London), 191.
2. Anne Savage, transl. and coll., *The Anglo-Saxon Chronicles* (New York: St. Martin's/Marek, 1983), 30, 84–87.
3. Patrick E. McGovern, *Uncorking the Past: The Quest for Wine, Beer, and Other Alcoholic Beverages* (Berkeley, Los Angeles, and London: University of California Press, 2009), 39.
4. Ibid., 143.
5. Ibid., 144–46.
6. Michael J. Enright, *Lady with a Mead Cup: Ritual, Prophecy and Lordship in the European Warband from La Tène to the Viking Age* (Dublin: Four Courts Press, 1996), 101.
7. McGovern, *Uncorking the Past*, 151.
8. Ransome, *Sacred Bee*, 195.
9. Ibid.
10. Tom Quinn Kumpf, *Ireland: Standing Stones to Stormont* (Boulder, CO: Devenish Press, 2004), http://www.knowth.com/teamhair.htm.
11. Ransome, *Sacred Bee*, 197.
12. Stephen Pollington, *The Mead Hall: The Feasting Tradition in Anglo-Saxon England* (Norfolk, UK: Anglo-Saxon Books, 1995), 130–38.
13. Ibid., 159.
14. Ann Hagen, *A Second Handbook of Anglo-Saxon Food and Drink: Production and Distribution* (Frithgarth, Thetford Forest Park, UK: Anglo-Saxon Books, 1995; reprinted 1999), 243.
15. Ibid., 204–5.

16. Pollington, *Mead Hall*, 31.

17. Ralph Metzner, *The Well of Remembrance: Rediscovering the Earth Wisdom Myths of Northern Europe* (Boston and London: Shambahala, 1994), 279–80.

18. Ibid., 283.

19. The Green Earth Foundation, http://www.greenearthfound.org/ralph_metzner.html.

20. McGovern, *Uncorking the Past*, 140–41.

21. Odd Nordland, *Brewing and Beer Traditions in Norway: The Social Anthropological Background of the Brewing Industry* (Gjovik: Mariendals Boktrykkeri, 1969), 216.

22. Stephen Harrod Buhner, *Sacred and Herbal Healing Beers*, (Boulder: Brewers Publications, 1998), 177–78.

23. Nordland, *Brewing and Beer Traditions in Norway*, 217.

24. Ibid.

25. Ibid.

26. Ibid.

27. Ibid., 218.

28. Buhner, *Sacred and Herbal Healing Beers*, 184.

29. James Green, *The Herbal Medicine-Maker's Handbook: A Home Manual* (Berkley: The Crossing Press, A Division of Ten Speed Press, 2000), 35.

30. Buhner, *Sacred and Herbal Healing Beers*, 185.

31. Ibid., 194.

32. Green, *The Herbal Medicine-Maker's Handbook*, 94.

33. Buhner, *Sacred and Herbal Healing Beers*, 190.

34. Ibid., 196.

35. Green, *The Herbal Medicine-Maker's Handbook*, 33.

36. Nordland, *Brewing and Beer Traditions in Norway*, 218.

37. http://www.spruceontap.com/aboutus.sc.

38. Claude Lévi-Strauss, *From Honey to Ashes: Introduction to a Science of Mythology: 2*, Transl. John and Doreen Weightman (New York, Evanston & San Francisco: Harper & Row Publishers, 1966; Transl. 1973), 67–68.

39. Buhner, *Sacred and Herbal Healing Beers*, 32.

40. Bo Almqvist, *Viking Ale: Studies on Folklore Contacts Between the Northern and the Western Worlds* (Aberystwyth, Wales: Boethius Press, 1991), 71.

41. Buhner, *Sacred and Herbal Healing Beers*, 169.

42. Metzner, *The Well of Remembrance*, 291.

43. Buhner, notes in *Sacred and Herbal Healing Beers*, 172n.

44. H. S. Corran, *History of Brewing* (North Pomfret, VT: David and Charles, 1975), 44–45.

45. Ibid., 45.

46. Ibid., 43.

47. Buhner, *Sacred and Herbal Healing Beers*, 172.

CHAPTER THREE: Honey and the Bees We Have to Thank for It

1. Buhner, *Sacred and Herbal Healing Beers*, 20.

2. Ransome, *Sacred Bee*, 20.

3. http://news.nationalgeographic.com/news/2006/10/061025-oldest-bee.html; http://www.nytimes.com/1995/05/23/science/which-came-first-bees-or-flowers-find-points-to-bees.html.

4. Ransome, *Sacred Bee*, 140.

5. Amos Ives Root et al., *The ABC & XYZ of Bee Culture*, 40th ed. (Medina, OH: A. I. Root Company, 1990), 190–91.

6. Tammy Horn, *Bees in America: How the Honey Bee Shaped a Nation* (Lexington, KY: University Press of Kentucky, 2005), 25–26.

7. Root et al., *ABC & XYZ of Bee Culture*, 195, 435.

8. Horn, *Bees in America*, 28.

9. Ibid., 28.

10. Root et al., *ABC & XYZ of Bee Culture*, 291.

11. Ibid., 195.

12. http://beeinformed.org/research-2.

13. https://www.youtube.com/watch?v=5DFKqgWuCBA.

14. http://www.bushfarms.com/beesfoursimplesteps.htm.

15. http://www.apiservices.com/articles/us/ktbh.htm.

16. Les Crowder and Heather Harrell, *Top-Bar Beekeeping: Organic Practices for Honeybee Health* (White River Junction, VT: Chelsea Green Publishing, 2012), 2.

17. Root et al., *ABC & XYZ of Bee Culture*, 206–7.

18. Private correspondence with the author.

19. Ransome, *Sacred Bee*, 19.

20. Ransome, *Sacred Bee*, 190.

21. Eimear Chaomhánach, *The Lore of the Bee, Its Keeper and Produce, in Irish and Other Folk Traditions* (University College Dublin, Department of Irish Folklore, 2002), http://www.ucd.ie/pages/99/articles/chaomh.pdf.

22. Peter C. Molan, "Why Honey Is Effective as a Medicine: Its Use in Modern Medicine," *Bee World* 80, no. 2 (1999): 81.

23. Peter B. Olaitan, Olufemi E. Adeleke, and Iyabo O. Ola, "Honey: A Reservoir for Microorganisms and an Inhibitory Agent for Microbes," *African Health Sciences* 7, no. 3 (September 2007), http://www.ncbi.nlm.nih.gov/pmc/articles/PMC2269714.

24. Gitte Laasby, "'Honey Laundering' Means Fake Honey Coming in from China, Experts Warn," *Milwaukee, Wisconsin, Journal Sentinel*, http://www.jsonline.com/blogs/news/206463151.html, May 9, 2013; Jesse Hirsch, "Honey Laundering: A Primer," *Modern Farmer*, http://modernfarmer.com/2013/05/honey-laundering-a-primer, May 5, 2013.

CHAPTER FOUR: Preparing for Battle

1. John Ehle, *The Cheeses and Wines of England and France* (South Deerfield, MA: New England Cheesemaking Supply, 1972), 201.

2. Ann Hagen, *A Second Handbook of Anglo-Saxon Food and Drink*, 219.

3. Sandor Ellix Katz, *The Art of Fermentation* (White River Junction, VT: Chelsea Green Publishing, 2012), 74.

CHAPTER FIVE: Brewing the Drink of the Gods

1. Buhner, *Sacred and Herbal Healing Beers*, 64–65.

2. Amy Stewart, *The Drunken Botanist: The Plants That Create the World's Great Drinks* (Chapel Hill, NC: Algonquin Books of Chapel Hill, 2013), 25.

3. Olaitan et al., "Honey: A Reservoir for Microorganisms," 159–65.

4. *Webster's New Universal Unabridged Dictionary*.

5. Jeff Sparrow, *Wild Brews: Beer Beyond the Influence of Brewer's Yeast* (Boulder, CO: Brewers Publications, 2005), 109.

6. Nordland, *Brewing and Beer Traditions in Norway*, 78.

7. Michael Jackson, "Odin's Glass of Nectar: Michael Jackson Learns the Secret of Norway's Home Brews, Passed Down Via Viking 'Magic Sticks,'" *The Independent*, http://www.beerhunter.com/documents/19133-000103.html (December 4, 1993; accessed February 28, 2015).

8. Nordland, *Brewing and Beer Traditions in Norway*, 79.

9. Katz, *Wild Fermentation*, 135.

CHAPTER SIX: Basic Mead Recipes and Some Variations

1. Ken Schramm, *The Compleat Meadmaker* (Boulder, CO: Brewers Publications, 2003), 161.

2. http://www.pitt.edu/~kloman/tej.html.

3. Katz, *Art of Fermentation*, 74.

CHAPTER SEVEN: Herbal, Vegetable, Floral, Fruit, and Cooking Meads

1. Stephen Harrod Buhner, *The Lost Language of Plants* (White River Junction, VT: Chelsea Green Publishing, 2002), 223–24.

2. William Owen Pughe, *A Dictionary of the Welsh Language.* [Preceded by] *A Grammar of the Welsh Language.* 2 vols. [in 3 pt.]. [Followed by] *An Outline of the Characteristics of the Welsh.* 2 vols. [in 4 pt.], 1832 (Google eBook, accessed February 16, 2015), 335.

3. Nordland, *Brewing and Beer Traditions in Norway*, 222.

4. Green, *Herbal Medicine-Maker's Handbook*, 37.

5. From *A Modern Herbal* by Maud Grieve, originally published in 1931 (online since 1995 at http://www.botanical.com/botanical/mgmh/h/horwhi33.html; accessed February 18, 2015).

6. Jesse Frost, *Bringing Wine Home*, Book 1 (Bugtussle, KY: Rough Draft Farmstead, 2013), 36.

7. Guerber, *Myths of the Norsemen*, 102.

8. Ibid., 103.

9. Ehle, *Cheeses and Wines of England and France*, 208.

10. Claude Jolicoeur, *The New Cider Maker's Handbook* (White River Junction, VT: Chelsea Green Publishing, 2013), 16.

11. Ehle, *Cheeses and Wines of England and France*, 209.

CHAPTER EIGHT: Bragots, Herbal Honey Beers, Grogs, and Other Oddities

1. Randy Mosher, *Radical Brewing: Recipes, Tales, and World-Altering Meditations in a Glass* (Boulder, CO: Brewers Publications, 2004), 274.

2. Ibid., 200.

3. Ibid.

4. Buhner, *Sacred and Herbal Healing Beers*, 152.

5. Ibid., 148–52.

6. Sparrow, *Wild Brews*, 129–30.

7. Mosher, *Radical Brewing*, 201.

8. Ibid.

9. Charles Herman Sulz, *A Treatise on Beverages; or, The Complete Practical Bottler* (New York: Dick & Fitzgerald Publishers, 1888), 813–14, http://chestofbooks.com/food/beverages/A-Treatise-On-Beverages/Horehound-Beer.html#.VS7nYtzF_y1.

CHAPTER NINE: How to Drink Like a Viking

1. W. H. Auden and P. B. Taylor, transl., *Hávamál (Sayings of Hár)* (New York: Random House, 1967), excerpted from Elder Edda, stanza 1 (http://sigewif.com/library/Havamal -Auden-Taylor.pdf).

2. Ibid., stanza 12.

3. McGovern, *Uncorking the Past*, 266.

4. Pollington, *Mead Hall*, 31.

5. Ibid., 33.

6. Ibid.

7. Enright, *Lady with a Mead Cup*, 2.

8. Ibid., 3–4; the source quoted is Fr. Klaeber, ed. *Beowulf and the Fight at Finnsburg* (1950). The Translation is that of Andre Crépin, "Wealhtheow's Offering of the Cup to Beowulf: A Study in Literary Structure" (1979), 44–58.

9. Ibid., 55.

10. http://www.hurstwic.org/history/articles/daily_living/text/games_and_sports.htm (accessed March 12, 2015).

11. http://www.vikinganswerlady.com/games.shtml (accessed March 12, 2015).

12. Theodore Murdock Andersson and Kari Ellen Gad Pages, trans. and ed., *Morkinskinna: The Earliest Icelandic Chronicle of the Norwegian Kings (1030–1157)* (displayed at Google Books by permission of Cornell University Press, Ithaca, NY; accessed March 12, 2015), 346.

13. Örnólfur Thorrson, ed., *The Sagas of Icelanders: A Selection* (New York: Viking Penguin, 2000), 75.

14. http://www.hurstwic.org/history/articles/daily_living/text/games_and_sports.htm (accessed March 12, 2015).

15. Terry, *Poems of the Elder Edda*, 5–8.

16. Ibid., 10.

TROUBLESHOOTING

1. http://blog.honest.com/what-is-potassium-sorbate.

2. Michael and Irene Ash, compilers, *Handbook of Green Chemicals* (Endicott, NY: Synapse Information Resources, 2004), 845–46.

❀ INDEX ❀

Note: Page numbers in *italics* refer to photographs and figures; page numbers followed by *t* refer to tables.

❧ ABOUT THE AUTHOR ❧

Jenna Zimmerman

Jereme Zimmerman grew up on his parents' northern Kentucky goat
farm, Twin Meadows, where he was also homeschooled. After gradu-
ating from Berea College in Berea, Kentucky, he moved to the Pacific
Northwest, where he immersed himself in the world of homebrewing.
As the world's only peace-loving, green-living Appalachian Yeti Viking,
Zimmerman writes and speaks regularly on fermentation, mead
making, homesteading, and good eating. He is a regular contributor to
various publications and websites, including *New Pioneer* and *Backwoods
Home* magazines. He writes for Earthineer.com as "RedHeadedYeti."
He currently lives in Berea with his wife, Jenna, and daughters, Sadie
and Maisie, where he practices urban homesteading and cavorts with
farmers, authors, and fellow sustainable-living enthusiasts.